RAND NATIONAL DEFENSE RESEARCH INSTITUTE

Integrating Active and Reserve Component Staff Organizations

Improving the Chances of Success

Laurinda L. Rohn, Agnes Gereben Schaefer, Gregory A. Schumacher,
Jennifer Kavanagh, Caroline Baxter, Amy Grace Donohue

Prepared for the Office of Secretary of Defense

For more information on this publication, visit www.rand.org/t/RR1869

Library of Congress Cataloging-in-Publication Data is available for this publication.

ISBN: 978-0-8330-9828-3

Published by the RAND Corporation, Santa Monica, Calif.

Preface

Separate active and reserve military organizations have existed since the founding of the nation, and efforts to integrate them more closely—for example, to achieve greater efficiency, to make standards and practices more consistent, or to ensure commonality of purpose—date back to at least 1947. Not all of these efforts have been successful. The research reported here examines the factors that could increase or decrease the likelihood of success in undertaking such integrations.

This report should be of interest to those concerned with active and reserve component organizational structure and cooperation. This research was sponsored by the Office of Reserve Integration within the Office of the Under Secretary of Defense for Personnel and Readiness and conducted within the Forces and Resources Policy Center of the RAND National Defense Research Institute, a federally funded research and development center sponsored by the Office of the Secretary of Defense, the Joint Staff, the Unified Combatant Commands, the Navy, the Marine Corps, the defense agencies, and the defense Intelligence Community.

For more information on the RAND Forces and Resources Policy Center, see www.rand.org/nsrd/ndri/centers/frp or contact the director (contact information is provided on the web page).

Contents

Figures and Tables

Figures

Tables

Summary

The existence of separate active and reserve components of the U.S. military dates back to the founding of the nation, and both components are mentioned in the Constitution.[1] Debates about the roles of the components, their relationships, and the appropriate degree of integration between them date back just as far.[2] Various groups and individuals, both within and outside of the federal government, have proposed merging or otherwise integrating active component (AC) and reserve component (RC) forces during recent decades, dating back to at least 1947.[3] The rationales for integration have ranged from saving

---

[1] See Constitution of the United States, Article I, Section 8, and Article II, Section 2.

[2] For a more detailed description of the history on this subject, see Rostker, Bernard, Charles Robert Roll Jr., Marney Peet, Marygail Brauner, Harry J. Thie, Roger Allen Brown, Glenn A. Gotz, Steve Drezner, Bruce W. Don, Ken Watman, Michael G. Shanley, Fred L. Frostic, Colin O. Halvorson, Norman T. O'Meara, Jeanne M. Jarvaise, Robert Howe, David A. Shlapak, William Schwabe, Adele Palmer, James H. Bigelow, Joseph G. Bolten, Deena Dizengoff, Jennifer H. Kawata, Hugh G. Massey, Robert Petruschell, Craig Moore, Thomas F. Lippiatt, Ronald E. Sortor, J. Michael Polich, David W. Grissmer, Sheila Nataraj Kirby, and Richard Buddin, *Assessing the Structure and Mix of Future Active and Reserve Forces: Final Report to the Secretary of Defense*, Santa Monica, Calif.: RAND Corporation, MR-140-1-OSD, 1992, pp. 15–37.

[3] According to the U.S. Department of Defense (DoD), the reserve component "consists of the Army National Guard of the United States, the Army Reserve, the Navy Reserve, the Marine Corps Reserve, the Air National Guard of the United States, the Air Force Reserve, and the Coast Guard Reserve" (Joint Chiefs of Staff, *DoD Dictionary of Military and Associated Terms*, Joint Publication 1-02, November 8, 2010, as amended through February 15, 2016, p. 203). Title 10 of the U.S. Code (Armed Services) calls each of those seven organizations a reserve component, with all seven collectively being called the reserve components (U.S. Code, Title 10, Armed Forces, Subtitle E, Reserve Components, Part I, Organization

money or reducing duplicative command structures to improving RC readiness and equipment.[4]

DoD currently operates under the Total Force policy, which dates back to the 1960s. More recently, DoD has described the Total Force policy as having "two principal tenets: to plan for the integrated use of all forces that are available—active, reserve, civilian, and allied (including host nation support)—and to use reserve forces as the primary augmentation for active forces" with the objective of providing "maximum military capability within fiscal constraints by integrating the capabilities and strengths of active and reserve units in the most cost-effective manner."[5]

Efforts to implement the Total Force policy through increased integration have focused primarily, and understandably, on the efficiency and effectiveness of operational units—those units that deploy and potentially engage in combat. Over time, however, integration efforts have evolved to include the integration of staff and headquarters organizations and other supporting organizations as well.

Many different factors could increase or decrease the likelihood of success in integrating AC and RC staff organizations. The primary purpose of this study was to examine those factors in order to develop insights that could improve the likelihood of success in future integration efforts. We took a two-part approach. First, we reviewed the

---

and Administration, Chapter 1003, Reserve Components Generally, Section 10101, Reserve Components Named). In this report, we use the DoD definition and refer to them collectively as the reserve component. We use the term *reserves* to refer to the Army, Navy, Air Force, Marine Corps, and Coast Guard Reserves.

Somewhat surprisingly, we could not find an official definition for the term *active component*. It is used in Joint Publication 1-02 but is not defined. The term is typically used in contrast to *reserve component* and includes the regular Army, Navy, Air Force, Marine Corps, and Coast Guard. People who join the regular forces serve full time (40 or more hours per week) in them. For this study, we use the term *active component* to refer to the five regular military services.

[4] For a description of many such proposals, see Federal Research Division, Library of Congress, *Historical Attempts to Reorganize the Reserve Components*, Washington, D.C., October 2007.

[5] Total Force Policy Study Group, *Total Force Policy Interim Report to the Congress*, U.S. Department of Defense, September 1990, p. 4.

management and organizational literature and previous studies focused on organizational integration, organizational change, and organizational success. Second, we studied cases in which active and reserve military organizations were integrated, to gain insights into the factors that might have been present in those cases and how those factors might have affected the results of the integration efforts.

Defining the Total Force and Total Force Integration

We first set out to find a single, official definition for the term *total force*. We found many different definitions, all of which included AC and RC military forces. Some definitions included only those forces, while others included various combinations of other elements, such as civilian personnel, contractors, foreign and retired personnel, and even allied military personnel.

Similarly, we found no single definition of *total force integration*. Although there seems to be agreement on the broad purpose of improving military effectiveness and resource efficiency, there is little agreement on what integration really means and how to tell when it has been achieved. The appropriate form for integration to take and what the end result will look like depend greatly on the goal of the integration in question. That specific goal is very important, as is ensuring that the stated goal be consistent with actions taken to achieve it.

Although no single form or goal of total force integration will necessarily apply in all cases, we suggest that the following statement could define the goal of total force integration in the ideal:

> Total force integration would be achieved if military units, staffs, and other supporting organizations were designed from the start with a mix of full-time and part-time personnel, military and nonmilitary, that yields the maximum effective capacity at the lowest cost, with planning, command, and resourcing decisions made to benefit the total force rather than individual components of it.

Adopting a goal like this would mean that the perceived equities of individual organizations (such as the active Army, Navy Reserve, or Air National Guard) would not be considered or protected but would give way to the benefit of the overall total force. Costs and benefits of full-time and part-time personnel, both military and civilian, would be considered before deciding on the mix of personnel that would most effectively and efficiently fulfill the needs of combatant commands and the nation.

Practices and Recommendations for Improving the Chances of Success in Future Integration Efforts

The review of the organizational and management literature that focused on organizational integration, organizational change, and organizational success yielded many factors that can affect the success of efforts to integrate organizations. We distilled these down to a list of "best practices" suggested by the literature.

In the case studies of AC/RC integration that we examined, these practices were present to varying degrees, and we describe those findings in this section. Two things should be kept in mind. First, when examining the case studies, we did not always find enough information to determine the extent to which a practice was followed. Second and most importantly, we present this information in case it can help future AC/RC integration efforts to improve the chances of success. We do not intend to judge how well any service has carried out its integration efforts or whether one service has integrated better than another.

The case studies revealed some additional best practices that are more specific to integrating AC and RC organizations. We also include those practices in this section.

All of the best practices described in this report are found to be important in the literature, our case studies, or both. But a subset of the practices stood out to the team as especially relevant to AC/RC staff integrations. We believe that, by focusing on this subset of practices, leaders can increase the chances of future integrations succeeding. We have included recommendations in this section associated with those practices.

Establish the Need and the Vision for Change

Clarifying why an organization needs to integrate or change and defining the organization's future end state and goals—the "vision"—is key to success. The U.S. Coast Guard and the U.S. Navy both had clear visions for integration, and that clarity provided benchmarks and metrics for success. For example, the Navy's vision was to manage all Navy personnel as an integrated, holistic force in order to accomplish the priorities of the operational fleet.

The Air Force and Army cases also reinforce the need for a clear vision for change. The Air Force is moving forward developing plans and testing various forms of organizational structures, complete with goals and measures, but has not yet developed a clear vision for what its total force will look like and how to implement it. Like the Air Force, the Army is pursuing new approaches to integration but lacks a vision with a clearly defined end state. As with the Air Force, there is general acceptance in the Army that different force-mix ratios are likely required to balance current readiness with future modernization, but clarity is lacking about the form that should take and what degree of integration is both desirable and achievable.

Recommendation: Articulate the Need for Change and Adopt a Clear Vision for the Integration

Changing an organization, especially a large one, is difficult. People tend to be more willing to support change if they understand why it is needed and what it will look like. A clear vision reduces uncertainty and provides a way to assess progress. Achieving success is much easier if success is well defined. It is also important that the vision for the integration be realistic and achievable.

Create a Coalition to Support the Change

Successfully implementing proposed organizational changes requires the support of both internal and external stakeholders. Our case studies provided no evidence that any of the services created a coalition to support cross-component integration efforts. Because senior leaders directed some of the integrations (particularly in the cases of the Coast Guard and the Navy), this is not altogether surprising. In other cases, such as the Air Force, Army, and Marine Corps, integration efforts evolved over time.

Communicate the Vision

Leaders need to communicate the vision for change within and outside the organization not just once but throughout the process. Our case-study findings reinforce the importance of this best practice. The Coast Guard communicated its vision for its 1994 integration effort in multiple ways. Other services, including the Air Force, have communicated their visions through service policies and regulations. The Marine Corps and Navy have sustained their efforts to communicate their vision for integration over time, whereas the Army has not so far sustained its efforts to develop, articulate, and communicate its vision.

Recommendation: Regularly Communicate the Vision for the Integration

Integration, like other forms of organizational change, is a process rather than an event. The vision for integration needs to be communicated early in the integration, but the communication needs to continue throughout the process.

Develop an Implementation Strategy, Including Goals and Measures

The literature emphasizes that a well-designed strategy that clarifies goals, assigns responsibility and accountability, identifies risks, and outlines mitigation strategies can streamline the implementation process and help maintain momentum.

Recommendation: Develop a Strategy for Implementing the Integration That Includes Clear Goals and Measures of Success

The literature emphasizes the importance of a well-designed implementation strategy that states goals, assigns responsibility, identifies risks and mitigation strategies, and describes measures of progress and success for those goals. A strategy that has clear goals and concrete ways to measure progress toward them is required to assess whether integration efforts are making progress and to identify problems quickly and adjust course accordingly. This can save both time and money during implementation and can increase the odds that the changes can be sustained.

Link the Vision and the Structure

The structure of an integrated organization—including not only the management reporting structure but also the functions undertaken by the organization and the people and groups represented in it—should be linked and consistent with the vision for the new organization. Our case studies indicate that this best practice is one of the critical potential failure points in the successful implementation of integration efforts. The Air Force, Coast Guard, Marine Corps, and Navy provide examples in which the vision for integration was articulated and then reinforced through structural and organizational alignment and staff processes. This has occurred in different ways, including aligning different organizational structures for associated units with the vision for integration, changing leadership positions to align with the vision, and changing service processes to mesh with the vision.

Recommendation: Ensure That the Planned Organizational Structure Is Consistent with the Vision for the Integration

Both the literature and our case studies reinforce the importance of this practice. Making the organizational structure—including not only the management reporting structure but also the functions the organization performs, the representation of the components in the organization, and even the organization's processes—consistent with the vision for the integration will help to ensure that the integration is sustainable.

Embed the Changes in the New Culture

Institutionalizing the changes in the organization's culture and including elements indicative of healthy organizations, such as flexibility, adaptability, and a focus on the quality of the organization's products, will help to ensure that the changes endure and that the new organization succeeds. Our case studies also reflect the importance of embedding changes in the new culture. The Coast Guard, the Marine Corps, and the Navy have institutionalized their integration efforts in new total force cultures, as well as in service policies and procedures.

Manage the Integration of Cultures

Combining cultures can be challenging and requires attention and intentional management. Creating an integration team is one way to do so. The Coast Guard and Marine Corps have both managed their culture integrations and have been able to significantly reduce the competition and animosity between the components.

Recommendation: Work to Develop a Total Force Culture in the Integrated Organization

Similarities between the active and reserve military cultures can make it easy to underestimate the cultural challenges in AC/RC integrations. The real cultural challenge seems to be developing a total force culture in the integrated organization that considers the welfare of the total force first, rather than the welfare of the individual components.

Maintain Momentum

Providing sufficient resources and institutionalizing organizational changes in policy and procedures can help to sustain momentum beyond the initial phases of change. Our case studies indicate that the Air Force, Coast Guard, Marine Corps, and Navy have all been able to maintain the momentum of their integration efforts by institutionalizing the change and embedding it in the services' cultures and organizational structures. The Army has not maintained the momentum of its integration efforts. Instead, they have evolved periodically, and the vision for those efforts has changed.

Remember the Importance of People

Resistance can sabotage even the best-intentioned change, and treating people well and fairly and empowering them during periods of change can help to reduce that resistance. Although the literature emphasizes this best practice, we did not find evidence in our case studies that the services either intentionally focused on this practice or intentionally ignored it.

Assess Progress and Adjust Accordingly

Determining appropriate measures of effectiveness, collecting and evaluating relevant data, and adjusting direction based on that evaluation can improve the chances of an integration succeeding. The importance of this practice is reinforced in the findings from our case studies. In the Marine Corps case, assessing progress and making adjustments has resulted in several incremental changes since 1925. These changes ultimately led to a level of integration between the AC and RC that is arguably one of the highest among the services.

Establish Unity of Command

The case studies highlight the importance of establishing unity of command in organizations that integrate AC and RC, even though it can be a significant challenge because of statutory and organizational structures. The Coast Guard, Marine Corps, and Navy have been able to establish unity of command in their integrated organizations. The Army and the Air Force are limited in this regard because of legal constraints associated with Title 10 and Title 32.[6]

Recommendation: Establish Unity of Command to the Greatest Extent Possible in the Integrated Organization

A single, well-defined chain of command is a hallmark of military organizations. Integrating AC and RC organizations can pose some major difficulties in this area, most particularly when integrating Title 10 and Title 32 service members. Although the existing legal constraints are unlikely to change much in the near term, the cases also revealed some interesting approaches that can increase unity of command within those constraints.

Address Statutory Barriers

Existing statutes limit the degree of integration and unity of command that can be achieved. There are constraints on the duties that RC members can perform in full-time and part-time roles. The Air Force

---

[6] *Title 10* refers to Title 10 of the U.S. Code, Armed Forces. *Title 32* refers to Title 32 of the U.S. Code, National Guard.

and the Army face additional statutory constraints as the only services with Title 32 National Guard organizations. Both the Air Force and the Army cases indicate that some integration is possible within the legal constraints, but it is more difficult to achieve and more limited in nature. Work-arounds do provide some flexibility.

Recommendation: Explicitly Consider Statutory Barriers and Potential Work-Arounds

Title 10 and Title 32 limit the degree of integration and unity of command that can be achieved for the Air Force and the Army. Other statutes limit the functions that RC members can perform. But some work-arounds do exist, and it is important to consider both the limitations and the potential work-arounds before undertaking an integration.

Colocate Active and Reserve Component Personnel in Integrated Organizations

The Army, Coast Guard, and Marine Corps cases highlight the importance of colocating integrated AC and RC units. Colocation allows combined training, which is particularly important for operational organizations.

Acknowledgments

We would like to thank the people from the Air Force, Army, Coast Guard, Marine Corps, and Navy who took the time to meet with us and share their expertise on component integration issues. We would especially like to thank A. T. Ryan, Office of Marine Corps Forces Reserve, for his assistance in scheduling multiple visits and interviews, and CAPT Edward A. Lizak, Navy Reserve Forces Command, and Dana C. Luton and Kristin Blake, U.S. Army Forces Command, for their extensive follow-up to requests for additional documents and clarifications.

From our sponsoring office, Matthew P. Dubois, Robert H. Smiley, CAPT Todd Heyne, and Col Lee Ackiss provided information and feedback during the course of the project.

Many of our RAND colleagues shared their expertise and contributed to our thinking, including Frank Camm, Stephen Dalzell, Michael H. Decker, Michael E. Linick, William Marcellino, Albert A. Robbert, Bernard D. Rostker, Peter Schirmer, Christopher M. Schnaubelt, Carra S. Sims, and Stephen Watts. LTC Heather Smigowski made significant intellectual contributions to the work and provided her perspective as an Army officer who has spent time in both the active and reserve components.

RAND Forces and Resources Policy Center director John D. Winkler and associate director Lisa M. Harrington contributed valuable guidance and helpful insights throughout the project. Phillip Padilla and Cole Sutera ably assisted our research, and Jerry M. Sollinger helped to make the report more readable. Al Robbert, Gary

Crone, and Craig Bond provided thoughtful and constructive reviews of the report and helped us to refine our thinking.

Any remaining errors are, of course, our responsibility.

Abbreviations

A1	Office of the Deputy Chief of Staff of the Air Force for Manpower, Personnel and Services
AC	active component
ACC	Air Combat Command (U.S. Air Force)
ADCON	administrative control
AFI	Air Force instruction (U.S. Air Force)
AFR	Air Force Reserve (U.S. Air Force)
AFRC	Air Force Reserve Command (U.S. Air Force)
AGR	Active Guard Reserve
ANG	Air National Guard (U.S. Air Force)
ARNG	Army National Guard (U.S. Army)
CAR	Chief of the Army Reserve (U.S. Army)
CG-00	commandant of the U.S. Coast Guard
CG-1	assistant commandant for human resources (U.S. Coast Guard)

CG-13	Reserve and Military Personnel Directorate (U.S. Coast Guard)
CNO	Chief of Naval Operations (U.S. Navy)
COMMARFORRES	commander of U.S. Marine Corps Forces Reserve
COMNAVRESFOR	commander, Navy Reserve Forces
DCMS	deputy commandant for mission support (U.S. Coast Guard)
DoD	U.S. Department of Defense
DUIC	derivative unit identification code
EOD	explosive ordnance disposal
FORSCOM	U.S. Army Forces Command
FTS	full-time support
GAO	U.S. General Accounting Office (now the U.S. Government Accountability Office)
HAF	Headquarters, Air Force
HHBn	headquarters and headquarters battalion
HQDA	Headquarters, Department of the Army
HQMC	Headquarters, Marine Corps
HR	human resources
I&I	inspector-instructor
IMA	individual mobilization augmentee
IRR	Individual Ready Reserve
i-Wing	integrated wing
JP	joint publication

MARFORRES	U.S. Marine Corps Forces Reserve
MAW	Marine aircraft wing (U.S. Marine Corps)
MCU	multicomponent unit
MOA	memorandum of agreement
MOU	memorandum of understanding
NAVRESFOR	Navy Reserve Forces (U.S. Navy)
NCFA	National Commission on the Future of the Army
NCSAF	National Commission on the Structure of the Air Force
OCAR	Office of the Chief of the Army Reserve
OPDIR	operational direction
RC	reserve component
RegAF	Regular Air Force
RTF	Realignment Task Force
SELRES	Selected Reserve (U.S. Navy)
TF2	Total Force Task Force
TFC	Total Force Continuum
UCMJ	Uniform Code of Military Justice
USAR	U.S. Army Reserve
USCGR	U.S. Coast Guard Reserve

CHAPTER ONE

Introduction

Background

The existence of separate active and reserve components of the U.S. military dates back to the founding of the nation, and both components are mentioned in the Constitution.[1] Debates about the roles of the components, their relationships, and the appropriate degree of integration between them date back just as far.[2] Various groups and individuals, both within and outside of the federal government, have proposed merging or otherwise integrating active component (AC) and reserve component (RC) forces during recent decades, dating back to at least 1947.[3] The rationales for integration have ranged from saving

---

[1] See Constitution of the United States, Article I, Section 8, and Article II, Section 2.

[2] For a more detailed description of the history on this subject, see Bernard Rostker, Charles Robert Roll Jr., Marney Peet, Marygail Brauner, Harry J. Thie, Roger Allen Brown, Glenn A. Gotz, Steve Drezner, Bruce W. Don, Ken Watman, Michael G. Shanley, Fred L. Frostic, Colin O. Halvorson, Norman T. O'Meara, Jeanne M. Jarvaise, Robert Howe, David A. Shlapak, William Schwabe, Adele Palmer, James H. Bigelow, Joseph G. Bolten, Deena Dizengoff, Jennifer H. Kawata, Hugh G. Massey, Robert Petruschell, Craig Moore, Thomas F. Lippiatt, Ronald E. Sortor, J. Michael Polich, David W. Grissmer, Sheila Nataraj Kirby, and Richard Buddin, *Assessing the Structure and Mix of Future Active and Reserve Forces: Final Report to the Secretary of Defense*, Santa Monica, Calif.: RAND Corporation, MR-140-1-OSD, 1992, pp. 15–37.

[3] According to the U.S. Department of Defense (DoD), the RC "consists of the Army National Guard of the United States, the Army Reserve, the Navy Reserve, the Marine Corps Reserve, the Air National Guard of the United States, the Air Force Reserve, and the Coast Guard Reserve" (Joint Chiefs of Staff, *DoD Dictionary of Military and Associated Terms*, Joint Publication [JP] 1-02, November 8, 2010, as amended through February 15,

money or reducing duplicative command structures to improving RC readiness and equipment.[4]

DoD currently operates under the Total Force policy, which dates back to the 1960s. The originator of the Total Force concept was Theodore C. Marrs, then the Deputy Assistant Secretary of the Air Force for Reserve Affairs, who sought to integrate the Air Reserves into the full mission set of the Air Force. With the help of a 1967 RAND study, Marrs, who had become the Deputy Assistant Secretary of Defense for Reserve Affairs, convinced Secretary of Defense Melvin Laird of the concept's DoD-wide applicability,[5] and, in 1970, Laird directed its adoption across the department. He articulated two important concepts. First, he wrote, "Emphasis will be given to the concurrent consideration of the total forces, active and reserve, to determine the most advantageous mix to support national strategy and meet the threat."[6] And second, "Guard and Reserve units and individuals of the Selected Reserves will be prepared to be the initial and primary source of augmentation of the active forces in any future emergency requiring a rapid and substantial expansion of the Active Forces."[7] In 1973, Laird's successor, James R. Schlesinger, declared, "The Total Force is no longer a 'concept.' It is now the Total Force Policy which integrates the Active,

---

2016, p. 203). Title 10 of the U.S. Code (Armed Services) calls each of those seven organizations an RC, with all seven collectively being called the RCs (10 U.S.C. § 10101). In this report, we use the DoD definition and refer to them collectively as the RC. We use the term reserves to refer to the Army, Navy, Air Force, Marine Corps, and Coast Guard Reserves.

Somewhat surprisingly, we could not find an official definition for the term *AC*. It is used in JP 1-02 but is not defined. The term is typically used in contrast to *RC* and includes the regular Army, Navy, Air Force, Marine Corps, and Coast Guard. People who join the regular forces serve full time (40 or more hours per week) in them. For this study, we use the term *AC* to refer to the five regular military services.

[4] For a description of many such proposals, see Federal Research Division, Library of Congress, *Historical Attempts to Reorganize the Reserve Components*, Washington, D.C., October 2007.

[5] John T. Correll, "Origins of the Total Force," *Air Force Magazine*, February 2011, pp. 94–97, p. 96.

[6] As quoted in Rostker et al., 1992, p. 33.

[7] As quoted in Rostker et al., 1992, p. 33.

Guard, and Reserve forces into a homogenous whole."[8] More recently, DoD has described the Total Force policy as having "two principal tenets: to plan for the integrated use of all forces that are available—active, reserve, civilian, and allied (including host nation support)—and to use reserve forces as the primary augmentation for active forces" with the objective of providing "maximum military capability within fiscal constraints by integrating the capabilities and strengths of active and reserve units in the most cost-effective manner."[9]

Efforts to implement the Total Force policy through increased integration have focused primarily, and understandably, on the efficiency and effectiveness of operational units—those units that deploy and potentially engage in combat. Over time, however, integration efforts have evolved to include the integration of staff and headquarters organizations and other supporting organizations as well.

AC and RC organizations can differ in many important ways, including how members enter the organizations; how assignments are determined; when, where, and how training occurs; and how promotions are determined. There can also be more-fundamental differences, such as in culture and attitudes. These and other factors can affect the success of any integration effort.

Study Objective and Approach

The primary objective of this study was to examine the factors that could increase or decrease the likelihood of success in integrating AC and RC staff organizations. Along with that broad topic, two additional questions were of interest. First, what are some possible measures of effectiveness for a successful organization? Second, how can we ensure that the perspectives of AC and RC stakeholders continue to be represented in integrated organizations?

---

[8] As quoted in Rostker et al., 1992, pp. 33–34.

[9] Total Force Policy Study Group, *Total Force Policy Interim Report to the Congress*, U.S. Department of Defense, September 1990, p. 4.

We took a two-part approach to examining these issues. First, we undertook a review of the management and organizational literature and previous studies focused on three related areas: factors that could affect the success of integrating previously distinct organizations, factors that could affect implementing change in organizations, and factors that are associated with successful organizations. These factors could include the characteristics of people or organizations, approaches or practices used, barriers encountered, pitfalls to be avoided, and ways that barriers or pitfalls might be overcome.

Second, we examined cases in which active and reserve military organizations were integrated in order to gain insights into what factors, including those revealed in the literature, might have been present in those cases and how they might have affected the results of the integration efforts. We examined available secondary sources on the cases and engaged in structured discussions with about 35 subject-matter experts, organizational leaders, and other members of the organizations in question.[10] The subjects of these discussions included the vision and goals of the integration efforts, how the efforts were undertaken, the challenges encountered during the efforts, and the results of completed efforts. The depth and breadth of information available on the cases was limited by normal military and staff turnover, the passage of time, and the fact that written documentation on military organizational changes is not always extensive, retained, or accessible.

We combined the results of the literature review and the case studies to develop a set of best practices for integrating active and reserve staff organizations and recommendations associated with some of those practices that are designed to improve the odds of success in future integration efforts.

---

[10] The RAND Human Subjects Protection Committee approved all methods, procedures, and instruments used in the study.

Organization of the Report

Chapter Two examines the concepts of a total force and total force integration to describe the context in which AC and RC organizational integration takes place. Chapter Three presents the results of our review of the literature on factors affecting organizational integration, change, and success. We also examined past integration efforts undertaken by the U.S. Air Force, U.S. Army, U.S. Coast Guard, U.S. Marine Corps, and U.S. Navy, and Chapters Four through Eight, respectively, document these case studies. Chapter Nine describes the findings and offers recommendations resulting from our literature reviews and case studies.

CHAPTER TWO

The Total Force and Total Force Integration

Because DoD efforts to integrate AC and RC organizations are currently carried out under the rubric of the Total Force policy, our first task was to look for the meanings of *total force* and *total force integration*. Was there a single, official definition of either term that would help to clarify how to integrate organizations and when the goal of integration has been achieved? This chapter describes the results of that search. The first section examines definitions of the total force. The second section examines the meaning of total force integration and suggests a goal for total force integration.

Defining the Total Force

Across DoD and in the broader defense community, *total force* is an often-discussed but somewhat ill-defined term. Although we found several definitions from various sources, both within and outside of DoD, we found no authoritative, agreed-upon definition. No definition appears in *DoD Dictionary of Military and Associated Terms*, nor does one appear in *Capstone Concept for Joint Operations: Joint Force 2020* or in *Doctrine for the Armed Forces of the United States*.[1] DoD Directive 5100.01, *Functions of the Department of Defense and Its Major*

[1] JP 1-02, 2010 (2016); Joint Chiefs of Staff, *Capstone Concept for Joint Operations: Joint Force 2020*, September 10, 2012; Joint Chiefs of Staff, *Doctrine for the Armed Forces of the United States*, JP 1, March 25, 2013.

Components, uses the term *Total Force management* but does not define what a total force is.[2]

All of the definitions found include AC and RC military forces but vary in the range of what is included beyond that. Some definitions include only the uniformed forces. For instance, a Secretary of the Army memorandum on the Army Total Force Policy directive states, "As one Total Force, the Active Army, Army National Guard and U.S. Army Reserve provide operating and generating forces to support the National Military Strategy and Army commitments worldwide."[3] Similarly, "Air Force Guidance Memorandum to AFI 90-1001" states, "Together the components ([Active Duty, Air National Guard, Air Force Reserve]) form the Air Force's Total Force."[4] The National Commission on the Future of the Army (NCFA) calls "the Regular Army, the Army National Guard, and the Army Reserve" a "Total Force."[5]

Some definitions expand the concept to include civilian personnel. In *Navy's Total Force Vision for the 21st Century*, the Navy describes

---

2 Director of Administration and Management, *Functions of the Department of Defense and Its Major Components*, DoD Directive 5100.01, December 21, 2010, p. 11.

3 John M. McHugh, Secretary of the Army, "Army Directive 2012-08 (Army Total Force Policy)," memorandum for principal officials of Headquarters, Department of the Army; commanders, U.S. Army Forces Command, U.S. Army Training and Doctrine Command, U.S. Army Materiel Command, U.S. Army Europe, U.S. Army Central, U.S. Army North, U.S. Army South, U.S. Army Pacific, U.S. Army Africa, U.S. Army Special Operations Command, Military Surface Deployment and Distribution Command, U.S. Army Space and Missile Defense Command/Army Strategic Command, U.S. Army Network Enterprise Technology Command/9th Signal Command (Army), U.S. Army Medical Command, U.S. Army Intelligence and Security Command, U.S. Army Criminal Investigation Command, U.S. Army Corps of Engineers, U.S. Army Military District of Washington, U.S. Army Test and Evaluation Command, and U.S. Army Installation Management Command; superintendent, U.S. Military Academy; and director, U.S. Army Acquisition Support Center, Washington, D.C.: U.S. Department of Army, September 4, 2012, p. 1.

4 Michael R. Moeller, deputy chief of staff, Strategic Plans and Programs, U.S. Air Force, "Air Force Guidance Memorandum to AFI 90-1001," Washington, D.C., Air Force Guidance Memorandum 01 to Air Force Instruction 90-1001, January 23, 2014, p. 3.

5 NCFA, *Report to the President and the Congress of the United States*, January 28, 2016, p. 9. Congress established the NCFA—with members appointed by Congress and by the President—to examine the size and structure of the Army.

"our Total Force, active and reserve Sailors and Navy civilians."[6] The National Commission on the Structure of the Air Force (NCSAF), like the Navy, includes "Active, Reserve, Guard, and civilian" personnel in its definition of a Total Force.[7]

Other definitions broaden even further to include contractors and foreign and retired personnel. The Reserve Forces Policy Board and a 1979 U.S. General Accounting Office (GAO) report both include contractors in addition to AC, civilian, and RC personnel in their definitions.[8] Similarly, the Office of the Assistant Secretary of Defense for Manpower and Reserve Affairs enumerates the parts of the total force as "the Active and Reserve components, the civilian work force, contracted support services, and host nation support" in its *Defense Manpower Requirements Report: Fiscal Year* 2017.[9] AFI 90-1001 says that the Air Force

> Total Force includes Regular Air Force, Air National Guard of the United States, and Air Force Reserve military personnel, US Air Force military retired members, US Air Force civilian personnel (including foreign national direct- and indirect-hire, as well as non-appropriated fund employees), contractor staff, and host-nation support personnel.[10]

---

[6] Chief of Naval Operations, *Navy's Total Force Vision for the 21st Century*, Washington, D.C.: U.S. Department of the Navy, January 2010, p. 3.

[7] NCSAF, *Report to the President and Congress of the United States*, Washington, D.C., January 30, 2014, p. 13. NCSAF was a congressionally mandated commission established to examine whether to restructure the U.S. Air Force and, if so, how.

[8] Reserve Forces Policy Board, *Eliminating Major Gaps in DoD Data on the Fully-Burdened and Life-Cycle Cost of Military Personnel: Cost Elements Should Be Mandated by Policy—Final Report to the Secretary of Defense*, U.S. Department of Defense, Report FY13-02, January 7, 2013, p. 8; GAO, *Comptroller General's Annual Report 1979*, Washington, D.C., B-119600, January 25, 1980, p. i.

[9] Office of the Assistant Secretary of Defense for Manpower and Reserve Affairs, *Total Force Planning and Requirements Directorate, Defense Manpower Requirements Report: Fiscal Year 2017*, Washington, D.C.: U.S. Department of Defense, April 2016, p. vi.

[10] AFI 90-1001, p. 37.

This definition, which differs from the one in the associated "Air Force Guidance Memorandum to AFI 90-1001" noted above, highlights an instance in which a single organization (in this case, the Air Force) can define *total force* differently in different places.

Secretary of Defense Laird defined the term quite broadly, even including allied military personnel:

> In defense planning, the Strategy of Realistic Deterrence emphasizes our need to plan for optimum use of all military and related resources available to meet the requirements of Free World security. These Free World military and related resources—which we call "Total Force"—include both active and reserve components of the U.S., those of our allies, and the additional military capabilities of our allies and friends that will be made available through local efforts, or through provision of appropriate security assistance programs.[11]

A somewhat different expansion to include units and equipment is provided in Marine Corps Order 5311.1E, which says, "The Total Force is defined as all units, billets (Marine, United States Navy (USN), and civilian) and equipment resident in the active component (AC) and the reserve component (RC)."[12]

Although these definitions differ in scope, they generally agree on the high-level objective of a total force: to fulfill DoD missions and to meet the requirements of the national security strategy and national military strategy and the needs of the combatant commanders.[13] It is also worth noting that definitions might differ to appeal to different audiences or to serve different purposes. For example, including allied

---

[11] Melvin R. Laird, Secretary of Defense, *National Security Strategy of Realistic Deterrence: Secretary of Defense Melvin R. Laird's Annual Defense Department Report FY 1973*, February 22, 1972, p. 9.

[12] Commandant of the Marine Corps, *Total Force Structure Process*, Marine Corps Order 5311.1E, November 18, 2015, p. 1.

[13] In keeping with his expansion to include allied personnel, Laird expanded the objective to meeting "the requirements of Free World security" (Laird, 1972, p. 9).

military personnel could help to foster a sense of inclusion, while focusing solely on military personnel could help to provide a sense of unity.

Defining Total Force Integration

As with the definition of a total force, there seems to be agreement on the broad purpose of total force integration, which is to provide "maximum military capability within fiscal constraints by integrating the capabilities and strengths of active and reserve units in the most cost-effective manner."[14] Although there might be agreement that the general purpose of integration is to improve military effectiveness and resource efficiency, there is little agreement on what integration really means and how to tell when it has been achieved. Total force integration is not well defined. In the context of military forces, there is a broad range of possible meanings or goals of integration. Figure 2.1 shows some examples.

The appropriate form for integration to take and what the end result will look like depend greatly on the goal of the integration. That specific goal is very important, as is ensuring that the stated goal be consistent with actions taken to achieve it.

Figure 2.1
Range of Possible Meanings of Integration

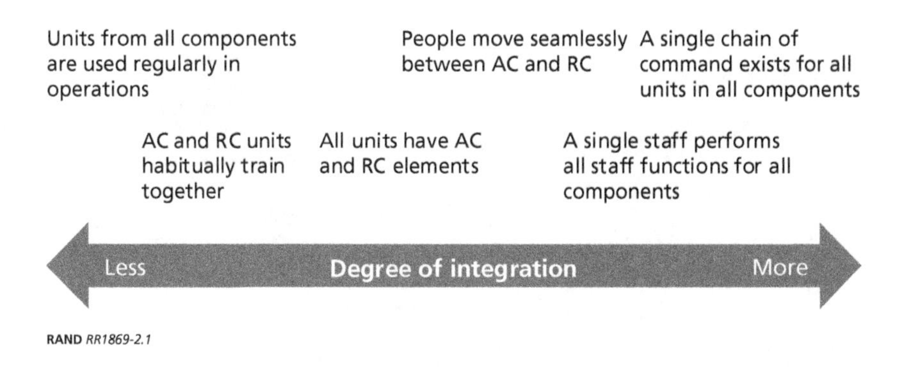

RAND *RR1869-2.1*

[14] Total Force Policy Study Group, 1990, p. 4.

Although no single form or goal of total force integration will necessarily apply in all cases, we suggest that the following statement could define the goal of total force integration in the ideal:

> Total force integration would be achieved if military units, staffs, and other supporting organizations were designed from the start with a mix of full-time and part-time personnel, military and nonmilitary, that yields the maximum effective capacity at a given cost with planning, command, and resourcing decisions made to benefit the total force rather than individual components of it.

Adopting a goal like this would mean that the perceived equities of individual organizations (such as the active Army, Navy Reserve, or Air National Guard) would not be considered or protected but would give way to the benefit of the overall total force. Costs and benefits of having full-time and part-time personnel, both military and civilian, would be considered before deciding on the mix of personnel that would most effectively and efficiently fulfill the needs of combatant commands and the nation.

CHAPTER THREE

Integrating and Changing Organizations Successfully

The primary objective of this study, as noted in Chapter One, was to examine the factors that could increase or decrease the likelihood of success in integrating AC and RC staff organizations. Types of factors that could affect the success of an integration—positively or negatively—might include the characteristics of the organizations being integrated, the characteristics of people within the organizations, approaches and practices for undertaking and implementing integrations, potential barriers to integration, ways to overcome potential barriers, and pitfalls to be avoided.

To see what insights it might offer on such factors, we surveyed the organizational and management literature that focuses on integrating organizations. Much of this subset of the literature derives from combining existing commercial organizations (for example, through mergers or acquisitions), and some covers the integration of minority groups (based on ethnicity or gender, for example) into organizations. In the first section of this chapter, we describe the results of this literature survey.

Integrating organizations is a specific type of organizational change. Therefore, the team also surveyed the organizational change literature to see which factors described in that broader literature might also apply to integration, and we describe those results in the second section of this chapter.

When undertaking organizational integration and change, improving or maintaining the success of the resulting organization is an underlying, if sometimes implicit, consideration. So the team also surveyed the literature on this topic both to see whether additional factors described in it might apply to organizational integration and to gain any insights on how to assess and perhaps measure organizational success. The third section of this chapter covers these results.

The fourth and final section of this chapter merges the factors discovered in the literature surveys and in the case studies to develop a framework for approaching and assessing efforts to integrate AC and RC staff organizations.

Factors That Affect the Success of Organizational Integrations

The literature on integrating organizations reveals several factors that could affect whether such integrations succeed. Marc Epstein identifies five drivers of successful postmerger integration: (1) a coherent integration strategy that "reinforces that this is a 'merger of equals' rather than an acquisition"; (2) a strong integration team that has representatives from all of the integrating organizations and that is focused on the integration, especially on eliminating any culture clashes in the integrated organization; (3) communication from senior management that is "significant, constant, and consistent," that builds confidence in the integration purpose and process, that reinforces the purpose of the integration "with a tangible set of goals," and that addresses important issues, such as personnel retention and separation policies; (4) speed in implementing the integration, which will reduce uncertainty and instability; and (5) measures of success that are aligned with the strategy and vision of the integration.[1] Epstein also emphasizes that critical choices on such matters as postintegration organizational structure, systems, processes, and practices "should not be made on the basis of imitating the status quo from one organization or the other" but rather "on the basis of a neutral, objective decision-making process that considers the solutions employed in the previous organization, as well as

[1] Marc J. Epstein, "The Drivers of Success in Post-Merger Integration," *Organizational Dynamics*, Vol. 33, Issue 2, May 2004, pp. 174–189, pp. 176–179.

any other alternatives."[2] He also notes that, "in personnel decisions, employees of both companies must be judged by the same standards and the candidate selection process based on merit rather than as a basis for a power struggle."[3]

De Noble, Gustafson, and Hergert identified eight lessons for postmerger integration success. First, focus on the source of integration problems rather than on the symptoms. The authors note that there is often pressure to complete a merger quickly, which creates an incentive to delay addressing potentially serious problems into the postmerger implementation phase.[4] Second, get line management involved early in the integration planning phase. Management can serve as a reality check on assumptions that planners make about such issues as new management structures or potential cost savings.[5] Third, cross-fertilize management teams. The authors point out that, "whenever a merger occurs, there is a psychological hurdle to surmount in establishing a new corporate identity. It is critical to replace the 'us' vs 'them' mentality with a spirit of teamwork."[6] Peter Drucker has also suggested that, during the first year of a merger, it is essential that a large number of people in the management groups of both companies receive substantial promotions across the lines—from one company to the other.[7] Fourth, people are very important. Treating people unfairly in layoffs, for example, can be demoralizing for those remaining, and "the loss of motivation and support can be disastrous."[8] Fifth, find the hidden costs. The authors warn that not considering costs for such activities as combining facilities and systems and relocating employees can lead to

---

[2] Epstein, 2004, p. 176.

[3] Epstein, 2004, p. 176.

[4] Alex F. De Noble, Loren T. Gustafson, and Michael Hergert, "Planning for Post-Merger Integration: Eight Lessons for Merger Success," *Long Range Planning*, Vol. 21, No. 4, August 1988, pp. 82–85, pp. 82–83.

[5] De Noble, Gustafson, and Hergert, 1988, p. 83.

[6] De Noble, Gustafson, and Hergert, 1988, p. 83.

[7] Peter Drucker, "The Five Rules of Successful Acquisition," *Wall Street Journal*, October 15, 1981, as cited in De Noble, Gustafson, and Hergert, 1988, p. 83.

[8] De Noble, Gustafson, and Hergert, 1988, pp. 83–84.

disappointing results.[9] Sixth, corporate culture will change. Integrating organizations have their own cultures, and failing to create a strategy for managing the process of combining them can undermine success.[10] Seventh, strategy and structure should be linked. The authors note that "an organization structure must properly reflect the underlying strategy of the organization in order to be successful" but that "[t]his simple concept is frequently overlooked."[11] Eighth, lessons should be applied earlier rather than later, so that problems are prevented. Waiting to address problems once they happen might be too late.[12]

Similarity in the cultures of the two integrating organizations is often cited as a factor crucial to success. In this context, *similarity* might mean sharing common values, goals, or standards and might include having similar processes or policies on such issues as staffing or expectations.[13]

Ensuring Minority Representation During Integration

As noted in Chapter One, one of the questions of special interest in this study was how to ensure that the perspectives of both AC and RC stakeholders can continue to be represented in an integrated organization. When two groups are integrated, the success of the integration and the experience of each group are likely to be affected by the proportions of each group—that is, how many members of each original group are included in the new organization. Literature on critical mass considers this issue and seeks to identify the number of members from a minority group that can safely and successfully be integrated into a majority without any sort of negative ramifications. Much of this

---

[9] De Noble, Gustafson, and Hergert, 1988, p. 84.

[10] De Noble, Gustafson, and Hergert, 1988, p. 84.

[11] De Noble, Gustafson, and Hergert, 1988, p. 84. The authors cite Alfred D. Chandler, *Strategy and Structure: Chapters in the History of the Industrial Enterprise*, Cambridge, Mass.: MIT Press, 1962.

[12] De Noble, Gustafson, and Hergert, 1988, pp. 84–85.

[13] Susan Cartwright and Cary L. Cooper, "The Role of Culture Compatibility in Successful Organizational Marriage," *Academy of Management Executive*, Vol. 7, No. 2, May 1993, pp. 57–70.

literature focuses on the success of small groups of women integrated into larger groups of men or small groups of a racial or ethnic minority integrated with majority-group members. Most of the literature on critical mass also concentrates on civilian contexts, including large corporations, corporate boards, and legislatures. However, the key observations from this literature can apply to integrating AC and RC staffs into a combined organization.

The literature on critical mass is clear that skewed groups—those with a large proportion of one group and a small minority of another—face performance degradation and dysfunctional group dynamics. An early study of gender dynamics suggested that skewed groups were affected by "the dynamics of tokenism," which are associated with reduced organizational performance, low morale among the minority group, and reduced cohesion.[14] This work suggests that the group dynamics would be improved if the representation of the minority group could be increased to 15 to 35 percent. This is a fairly wide range and might not be all that helpful in determining the appropriate representation ratios. The adverse effects of skewed groups have also been found in the construction profession, in which women make up less than 5 percent of the total workforce and, as a result, face occupational isolation and limited opportunities for professional development and promotion.[15]

Literature on minority–majority representation among ethnic groups similarly suggests that having sufficient representation of minority groups can influence performance, morale, and both individual and organizational success. For example, a study of community colleges found that, at schools with higher representation of Latinos among both the student body and the faculty, Latino students have higher

---

[14] Rosabeth Moss Kanter, "Some Effects of Proportions on Group Life: Skewed Sex Ratios and Responses to Token Women," *American Journal of Sociology*, Vol. 82, No. 5, March 1977, pp. 965–990.

[15] Clara Greed, "Women in the Construction Profession: Achieving Critical Mass," *Gender, Work and Organization*, Vol. 7, No. 3, July 2000, pp. 181–196.

academic achievement, English ability, and academic success.[16] Similarly, African American students on a predominantly white campus have demonstrated social isolation and lack of confidence that affected academic performance.[17]

However, there is little agreement within the literature about specific thresholds for what qualifies as a sufficient critical mass or about the ideal integration proportions most likely to contribute to successful organizational transformation. A 2008 study of women on Fortune 1000 company boards found that, on an average ten-person board, having at least three female members is optimal.[18] A study of five Scandinavian parliaments similarly identified 30 percent as the "tipping point" but also noted that such factors as the attitudes of group members play a significant role in determining the optimal proportion of minority- to majority-group members.[19] However, a 2001 study on women in the New Zealand Parliament between 1975 and 1999 found that, even at 30 percent, women could not alter the legislature's culture or policy decisions, although they were more actively involved in debates on key issues, such as parental leave.[20] Despite this disagreement, research suggests rather universally that the extreme case, that of a solo representative of a minority within a larger group, results in decreases in performance for the group and the individual and can

---

[16] Linda Serra Hagedorn, Winny Chi, Rita M. Cepeda, and Melissa McLain, "An Investigation of Critical Mass: The Role of Latino Representation in the Success of Urban Community College Students," *Research in Higher Education*, Vol. 48, No. 1, February 2007, pp. 73–91.

[17] Jacqueline Fleming, *Blacks in College: A Comparative Study of Students' Success in Black and in White Institutions*, San Francisco, Calif.: Jossey-Bass Publishers, 1984.

[18] Alison M. Konrad, Vicki Kramer, and Sumru Erkut, "Critical Mass: The Impact of Three or More Women on Corporate Boards," *Organizational Dynamics*, Vol. 37, No. 2, 2008, pp. 145–164.

[19] Drude Dahlerup, "From a Small to a Large Minority: Women in Scandinavian Politics," *Scandinavian Political Studies*, Vol. 11, No. 4, December 1988, pp. 275–298.

[20] Sandra Grey, *Does Size Matter? Critical Mass and Women MPs in New-Zealand House of Representatives*, paper written for the 51st Political Studies Association Conference, Manchester, UK, April 10–12, 2001.

affect the dynamics within the group as a whole.[21] Increasing the representation of the minority group can reduce this effect by adding social support and allowing minority group members to develop network connections.[22]

However, despite the evidence in support of critical-mass theory, there are many skeptics and those who argue that a focus on minority–majority proportions in a group has little relationship to group performance or group interactions.[23] Furthermore, some studies find little evidence for any sort of critical-mass effect. Specifically, a 2011 study of American corporate board members found that the women who were in the minority or even the only female on the board embraced their "pathbreaker" status and did not perceive performance degradation or other detrimental effects of the "skewed" nature of the corporate board.[24]

The literature on critical mass suggests a few key insights relevant to organizational mergers, such as the integration of active and reserve staffs. First, despite some skepticism, it does seem that the relative proportions of the two integrating organizations do have some effect on organizational and individual performance, as well as cohesion and

---

[21] Charles G. Lord and Delia S. Saenz, "Memory Deficits and Memory Surfeits: Differential Cognitive Consequences of Tokenism for Tokens and Observers," *Journal of Personality and Social Psychology*, Vol. 49, No. 4, 1985, pp. 918–926; Denise Sekaquaptewa and Mischa Thompson, "Solo Status, Stereotype Threat, and Performance Expectancies: Their Effects on Women's Performance," *Journal of Experimental Social Psychology*, Vol. 39, No. 1, January 2003, pp. 68–74.

[22] Laura Smart Richman, Michelle vanDellen, and Wendy Wood, "How Women Cope: Being a Numerical Minority in a Male-Dominated Profession," *Journal of Social Issues*, Vol. 67, No. 3, September 2011, pp. 492–509.

[23] See, for example, Paul Chaney, "Critical Mass, Deliberation and the Substantive Representation of Women: Evidence from the UK's Devolution Programme," *Political Studies*, Vol. 54, No. 4, December 2006, pp. 691–714; Amy Caiazza, "Does Women's Representation in Elected Office Lead to Women-Friendly Policy? Analysis of State-Level Data," *Women and Politics*, Vol. 26, No. 1, 2004, pp. 35–70; and Sarah Childs, Paul Webb, and Sally Marthaler, "Constituting and Substantively Representing Women: Applying New Approaches to a UK Case Study," *Politics and Gender*, Vol. 6, No. 2, June 2010, pp. 199–223.

[24] Lissa Lamkin Broome, John M. Conley, and Kimberly D. Krawiec, "Dangerous Categories: Narratives of Corporate Board Diversity," *North Carolina Law Review*, Vol. 89, 2011, pp. 760–808.

individual morale. Further, it seems that having a grossly unbalanced ratio can often also have detrimental effects on performance, success, morale, and group dynamics. However, it is far from certain what the "optimal" ratio is or what level of representation is needed to achieve sufficient critical mass. Studies range in their estimates from 15 to 35 percent, with some suggesting that even 35 percent is not enough to guarantee that the minority group will be able to influence group decisions and will not feel isolated or excluded. Robbert et al. suggest that RC representation of at least 20 percent "would be sufficient to ensure that the RC has effective voice in the total force."[25]

Finally, it is likely that other factors, such as the attitudes of majority-group members, the overall organizational culture, organizational leadership, and other contextual attributes, play a large role in determining the level of minority-group representation to support organizational effectiveness, particularly in situations of organizational mergers or integration. Above all, fostering an inclusive culture that values and treats and promotes members of both or all integrating groups equally, combined with other organizational success factors described elsewhere in this report, will probably be best suited to supporting effective organizational change and integration.

Factors That Affect the Success of Implementing Organizational Change

Integrating AC and RC staffs can be a major organizational change. In addition to surveying the literature on organizational integration, we surveyed the literature on the broader issue of organizational change to look for additional relevant insights into how to go about undertaking and measuring organizational change, barriers to organizational change and ways to overcome them, and errors to avoid in implementing organizational change.

---

[25] Al Robbert, William A. Williams, and Cynthia R. Cook, *Principles for Determining the Air Force Active/Reserve Mix*, Santa Monica, Calif.: RAND Corporation, MR-1091-AF, 1999, p. 23.

Models of Organizational Change

A variety of process models in the literature try to identify the different phases of organizational change.[26] In his book *A Force for Change*, John Kotter developed one of the most-prominent models.[27] Kotter's change-phase model consists of eight critical phases of change that should be implemented in the following sequence: (1) establish a sense of urgency, (2) create a coalition, (3) develop a clear vision, (4) share the vision, (5) empower people to clear obstacles, (6) secure short wins, (7) consolidate and keep moving, and (8) anchor the change in the organizational culture.[28] Cummings and Worley described a five-phase, general process for managing change that has a very similar structure: (1) motivate change, (2) create vision, (3) develop political support, (4) manage transition, and (5) sustain momentum.[29] Peter deLeon argued that policy innovation can be conceived as moving through six stages: (1) agenda setting, (2) adoption, (3) early implementation, (4) execution, (5) evaluation and modification, and (6) later implementation to completion.[30] Termination of the change process can occur at any point.

In addition to the change-phase models mentioned above, two other types of prescriptive models of organizational change—bottom-up models and top-down models—emerge from the literature. Bottom-

---

[26] See Richard Beckhard and Reuben T. Harris, *Organizational Transitions: Managing Complex Change*, 2nd ed., Reading, Mass.: Addison-Wesley, 1987; W. Warner Burke and George H. Litwin, "A Causal Model of Organizational Performance and Change," *Journal of Management*, Vol. 18, No. 3, 1992, pp. 523–545; Gary M. Grobman, "Complexity Theory: A New Way to Look at Organizational Change," *Public Administration Quarterly*, Vol. 29, No. 3–4, Fall 2005–Winter 2006, pp. 350–382; Andrew H. van de Ven and Marshall Scott Poole, "Explaining Development and Change in Organizations," *Academy of Management Review*, Vol. 20, No. 3, July 1995, pp. 510–540.

[27] John P. Kotter, *A Force for Change: How Leadership Differs from Management*, New York: Free Press, 1990.

[28] Kotter, 1990.

[29] Thomas G. Cummings and Christopher G. Worley, *Organization Development and Change*, 5th ed., St. Paul, Minn.: West Publishing, 1993.

[30] Peter deLeon, "The Stages Approach to the Policy Process: What Has It Done? Where Is It Going?" in Paul A. Sabatier, ed., *Theories of the Policy Process*, Boulder, Colo.: Westview Press, 1999, pp. 19–34.

up models of organizational change focus on the role that rank-and-file members of an organization play in bringing about organizational change.[31] The literature argues that bottom-up approaches can be more successful and easier to implement than other approaches—the assumption being that, because the ideas for change are generated from below, it will be easier for management to acquire the buy-in of the rank and file.[32] The emphasis on bottom-up theories of organizational change has evolved into an emphasis in the literature on decentralized organizations. Decentralized organizations, by definition, are flatter, less hierarchical organizations than centralized ones, and advocates of such organizations argue that they are more adaptive and responsive to changing environments than hierarchical organizations.[33] Because military organizations are typically centralized and hierarchical, bottom-up approaches might be less applicable in integrating higher-level staff organizations.

Unlike bottom-up models of organizational change that emphasize grassroots mobilization for change, top-down models of organizational change argue that successful organizational change is imposed from upper management down to the rank and file. One of the most-important strategies for implementing organizational change is to enlist the support of a high-level manager or "change agent" to fight for and protect efforts associated with organizational change.[34] Such change

---

[31] See Paul A. Sabatier, "Top-Down and Bottom-Up Approaches to Implementation Research: A Critical Analysis and Suggested Synthesis," *Journal of Public Policy*, Vol. 6, No. 1, January–March 1986, pp. 21–48; Michael Lipsky, *Street-Level Bureaucracy: Dilemmas of the Individual in Public Services*, New York: Russell Sage Foundation, 1980; Benny Hjern and David O. Porter, "Implementation Structures: A New Unit of Administrative Analysis," *Organization Studies*, Vol. 2, No. 1, 1981, pp. 211–227; David Mechanic, "Sources of Power of Lower Participants in Complex Organizations," *Administrative Science Quarterly*, Vol. 7, No. 3, December 1962, pp. 349–364; and Michael Moon, "Bottom-Up Instigated Organizational Change Through Constructionist Conversation," *Journal of Knowledge Management Practice*, Vol. 9, No. 4, December 2008.

[32] Sabatier, 1986.

[33] Ori Brafman and Rod A. Beckstrom, *The Starfish and the Spider: The Unstoppable Power of Leaderless Organizations*, New York: Portfolio, 2006.

[34] W. Henry Lambright, "Leadership and Change at NASA: Sean O'Keefe as Administrator," *Public Administration Review*, March–April 2008, pp. 230–240; Sergio Fernandez and

agents often spur efforts for transformative change in an organization, and the most important function of a change agent is to support and fight for organizational change.[35]

However, not all who set out to become change agents succeed in changing an organization. Fernandez and Rainey argued that successful change agents are particularly mindful of the following eight activities: (1) ensure the need for change, (2) provide a plan for change, (3) build internal support for change and overcome resistance, (4) ensure top-management support and commitment to the change, (5) build external support for the change, (6) provide resources for the change, (7) institutionalize change, and (8) pursue comprehensive change.[36]

Ways to Measure Organizational Change

Measuring and monitoring organizational change is critical to identifying whether the goals of the change are being met, to monitoring implementation, and to identifying problems during implementation so that they can be addressed quickly. *Measurement of progress should be considered during the planning phase before any changes are made. By thinking through the measures that will be used to gauge progress, the organization can collect necessary data prior to the implementation of a decision so that it can then make comparisons across data collected prior to and after the decision.* The literature identifies various ways to measure organizational change, including measuring progress at various levels of

---

Hal G. Rainey, "Managing Successful Organizational Change in the Public Sector," *Public Administration Review*, Vol. 66, No. 2, March 2006, pp. 168–176. A change agent is a person either within an organization or external to an organization who helps the organization transform itself.

[35] See Oswald Jones, "Developing Absorptive Capacity in Mature Organizations: The Change Agent's Role," *Management Learning*, Vol. 37, No. 3, 2006, pp. 355–376; Richard T. Pascale and Jerry Sternin, "Your Company's Secret Change Agents," *Harvard Business Review*, May 2005; and Luis Almeida Costa, João Amaro de Matos, and Miguel Pina e Cunha, "The Manager as Change Agent: Communication Channels, Timing of Information, and Attitude Change," *International Studies of Management and Organization*, Vol. 33, No. 4, Winter 2003–2004, pp. 65–93.

[36] Fernandez and Rainey, 2006.

the organization. For instance, these measurements could include individual employee assessments (e.g., proficiency metrics, employee feedback), as well as organizational assessments (e.g., tracking of change-management activities conducted according to plan).[37]

An implementation plan is also key to successful organizational change because it lays out the road map for implementation of the changes and identifies the steps necessary for implementing the changes. A fundamental aspect of this implementation plan should be a monitoring plan that identifies the types of issues to monitor (what are you measuring?), metrics for measuring those issues (how are you measuring progress, and what information do you need?), and methods for collecting data (how are you collecting the information that you need to measure progress?).[38] In addition, it is key to identify who in the organization will be accountable for each of the items in the monitoring plan.

Overcoming Barriers to Organizational Change

Efforts to change organizations often fail because either the organizational culture or people in the organization are resistant to change.[39] The broader literature identifies the following individual sources of resistance: fear of the unknown, self-interest, habit, personality conflicts, differing perceptions, general mistrust, and social disruptions. The broader literature also identifies the following organizational sources of resistance: structural inertia, bureaucratic inertia, group norms, a resistant organizational culture, threatened power, threatened expertise, and threatened resource allocation.

Agents of change have to find ways for the organizational culture to accept change as less frightening than stability. The literature on

---

[37] Timothy J. Creasey and Robert Stise, eds., *Best Practices in Change Management: 1120 Participants Share Lessons and Best Practices in Change Management*, 9th ed., Loveland, Colo.: Prosci, 2016.

[38] Agnes Gereben Schaefer, Jennie W. Wenger, Jennifer Kavanagh, Jonathan P. Wong, Gillian S. Oak, Thomas E. Trail, and Todd Nichols, *Implications of Integrating Women into the Marine Corps Infantry*, Santa Monica, Calif.: RAND Corporation, RR-1103-USMC, 2015.

[39] Beckhard and Harris, 1987.

organizational change identifies, among others, the following means to deal with resistance to change: gradualism, education and communication, participation and involvement, negotiation and agreement, burden sharing, manipulation and co-option, explicit and implicit coercion, divide and conquer, and buy-out.[40] Their work seen as a milestone in the field, Beckhard and Harris argued that all three of the following components must be present to overcome the resistance to change in an organization: dissatisfaction with the present situation, vision of what is possible in the future, and achievable first steps toward reaching this vision.[41]

Errors to Avoid in Implementing Organizational Change

The literature on lessons learned from the implementation of organizational change also offers guidance for implementing organizational change.[42] For instance, Carol Kinsey Goman identified nine of the biggest mistakes in managing organizational change: (1) not understanding the importance of people; (2) not appreciating that people throughout the organization have different reactions to change; (3) treating transformation as an event, rather than a mental, physical, and emotional process; (4) being less than candid; (5) not appropriately "setting the stage" for change; (6) trying to manage transformation with the same strategies used for incremental change; (7) forgetting to negotiate the new "compact" between employers and employees; (8) believing that change communication was what employees heard or read from corporate headquarters; and (9) underestimating human potential.[43]

---

[40] Daniel T. Holt, Achilles A. Armenakis, Hubert S. Feild, and Stanley G. Harris, "Readiness for Organizational Change: The Systematic Development of a Scale," *Journal of Applied Behavioral Science*, Vol. 43, No. 2, June 2007, pp. 232–255.

[41] Beckhard and Harris, 1987.

[42] Carol Kinsey Goman, "The Biggest Mistakes in Managing Change," *Innovative Leader*, Vol. 9, No. 12, December 2000; Bryne Purchase, "Strategies for Implementing Organizational Change in a Public Sector Context: The Case of Canada," *TDRI Quarterly Review*, Vol. 11, No. 4, December 1996, pp. 27–35.

[43] Goman, 2000.

In a derivative of his eight-stage model of organizational change, Kotter later offered the diagram in Figure 3.1 that outlines the eight errors common to organizational change efforts and their consequences.

Factors That Affect Organizational Success

The research literature, including that focused on public-sector organizations and that focused on private organizations, identifies factors that can contribute to organizational success. These factors are relevant to all organizations but might be particularly important following a merger or organizational change that affects the organization's structure, leadership, responsibilities, or personnel. We describe these factors briefly below.

One prevalent factor is the existence of a clear mission or vision that is consistent, can be easily articulated, and can be used to shape

Figure 3.1
Eight Errors Common to Organizational Change Efforts and Their Consequences

Common errors

- Allowing too much complacency
- Failing to create a sufficiently powerful guiding coalition
- Underestimating the power of vision
- Undercommunicating the vision by a power of 10 (or 100 or even 1,000)
- Permitting obstacles to block the new vision
- Failing to create short-term wins
- Declaring victory too soon
- Neglecting to anchor changes firmly in the corporate culture

↓

Consequences

- New strategies are not implemented well
- Acquisitions do not achieve expected synergies
- Reengineering takes too long and costs too much
- Quality programs do not deliver hoped-for results

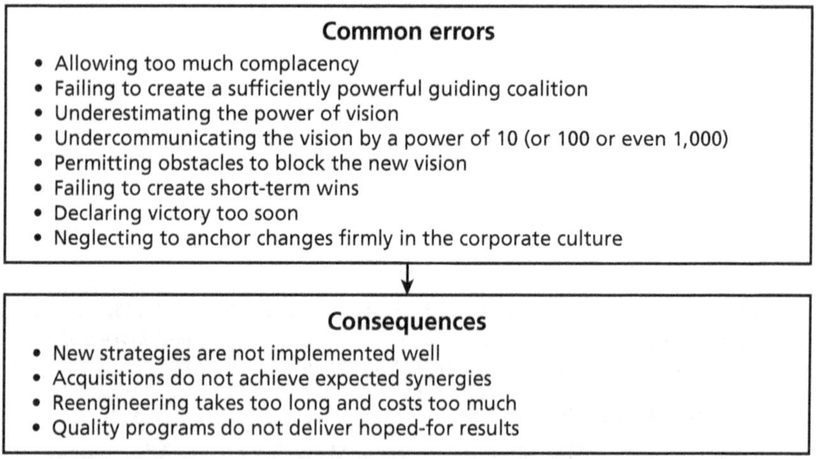

SOURCE: John P. Kotter, *Leading Change*, Boston, Mass.: Harvard Business School Press, 1996.

RAND *RR1869-3.1*

and guide the organization's day-to-day operations.[44] This vision provides coherence to the organization's decisions and serves as a motivator for personnel.[45] Strong leadership and managers' commitment to the organization's purpose or goal is another characteristic of successful organizations.[46] Strong leadership is required for a variety of reasons. It can prevent internal disagreements and rivalries, especially after a merger, acquisition, or other organizational change.[47] It can also help the organization stay focused on its vision and objectives and to set priorities that are consistent with the organization's vision or objectives.[48] Strong leadership can also be required to make difficult decisions about the organization's future. Management commitment is related to leadership but extends more broadly. Managers must serve not only as leaders but also as examples for the rest of the organization. They must be committed to and vocal champions of the organization's vision, values, and responsibilities, especially during times of organizational change.[49]

Leadership and organizational vision are closely related to another important characteristic of successful organizations: organizational

---

[44] Jia Wang, "Applying Western Organization Development in China: Lessons from a Case of Success," *Journal of European Industrial Training*, Vol. 34, No. 1, 2010, pp. 54–69; John C. Sawhill and David Williamson, "Mission Impossible? Measuring Success in Nonprofit Organizations," *Nonprofit Management and Leadership*, Vol. 11, No. 3, Spring 2001, pp. 371–386.

[45] Bernard Marr, *Managing and Delivering Performance: How Government, Public Sector, and Not-for-Profit Organizations Can Measure and Manage What Really Matters*, Amsterdam: Butterworth-Heinemann/Elsevier, 2009.

[46] Penny Gardiner and Peter Whiting, "Success Factors in Learning Organizations: An Empirical Study," *Industrial and Commercial Training*, Vol. 29, No. 2, 1997, pp. 41–48.

[47] Fernandez and Rainey, 2006.

[48] Gardiner and Whiting, 1997; Aldona Frączkiewicz-Wronka, Jacek Szołtysek, and Maria Kotas, "Key Success Factors of Social Services Organizations in the Public Sector," *Management*, Vol. 16, No. 2, December 2012, pp. 231–255; Kenneth J. Meier and Laurence J. O'Toole Jr., "Public Management and Organizational Performance: The Effect of Managerial Quality," *Journal of Policy Analysis and Management*, Vol. 21, No. 4, Autumn 2002, pp. 629–643.

[49] Meier and O'Toole, 2002; Gardiner and Whiting, 1997; Frączkiewicz-Wronka, Szołtysek, and Kotas, 2012; Fernandez and Rainey, 2006.

culture.[50] It is important that an organization's culture be consistent with its vision and objectives and that it promote innovation, efficiency, and high quality, as described in more detail below. Also important is having a culture that supports knowledge-sharing across departments and employees.[51] Of course, some cultures might be more adaptable and might adjust more easily to organizational change, a quality that can be important when the organization is integrating with other organizations, absorbing new staff, or undergoing other transformations. As a result, flexibility and adaptability can also be core characteristics of organizational success.[52]

Another important characteristic of successful organizations is having some system to promote, manage, and assess their operational processes. This includes a clear policy on quality or standards and a system by which quality can be measured and periodically assessed.[53] For organizations that produce intangible outputs, such as knowledge, leadership, guidance, or strategy (as is true in many organizations throughout DoD), quality can be more difficult to measure. However, successful organizations working in these areas create some kind of quality assurance process and instill the values associated with quality assurance in their employees. Successful organizations also focus on and promote measures, such as efficiency and innovation. This might mean efficiency in completing tasks or meeting deadlines but also includes innovation in developing new cutting-edge methodologies or

---

[50] Jeffrey Pfeffer and John F. Veiga, "Putting People First for Organizational Success," *Academy of Management Executive*, Vol. 13, No. 2, May 1999, pp. 37–48.

[51] Alawi, Adel Ismail al-, Nayla Yousif Al-Marzooqi, and Yasmeen Fraidoon Mohammed, "Organizational Culture and Knowledge Sharing: Critical Success Factors," *Journal of Knowledge Management*, Vol. 11, No. 2, 2007, pp. 22–42.

[52] Andrew M. Pettigrew, Richard W. Woodman, and Kim S. Cameron, "Studying Organizational Change and Development: Challenges for Future Research," *Academy of Management Journal*, Vol. 44, No. 4, August 2001, pp. 697–713.

[53] Andrea Rangone, "Linking Organizational Effectiveness, Key Success Factors and Performance Measures: An Analytical Framework," *Management Accounting Research*, Vol. 8, No. 2, June 1997, pp. 207–219.

ways of doing business.[54] A constant focus on quality, efficiency, and innovation are hallmarks of successful organizations and are particularly important during periods of flux or organizational change, when it can be easy for quality standards to be pushed aside or for a focus on consolidation to displace a focus on innovation and advancement.[55]

In addition to a focus on processes, successful organizations focus on their employees. This means not only emphasizing employee morale and retention but also promoting teamwork, training, and education.[56] Supporting employees with adequate training to complete their responsibilities and giving employees ample opportunities for development and advancement can be particularly important to encouraging employee morale and commitment to the organization.[57] Including employees in organizational decisions can also help encourage organizational loyalty and improve overall quality and efficiency.[58] This is especially important in cases of organizational change, in which employee involvement can help to reduce resistance to the change and improve the odds that any organizational transformation will succeed.[59] Strong leadership and having a flexible organizational culture can contribute to employee morale.[60] It is also important to ensure that the right types of employees are being hired to meet the organization's demands. This means that there needs to be a clear match between the types of people

---

[54] Rangone, 1997; Fariborz Damanpour, "Organizational Innovation: A Meta-Analysis of Effects of Determinants and Moderators," *Academy of Management Journal*, Vol. 34, No. 3, September 1991, pp. 555–590.

[55] Damanpour, 1991.

[56] Fernandez and Rainey, 2006; Pfeffer and Veiga, 1999; Benjamin Schneider, Sarah K. Gunnarson, and Kathryn Niles-Jolly, "Creating the Climate and Culture of Success," *Organizational Dynamics*, Vol. 23, No. 1, Summer 1994, pp. 17–29.

[57] Bruce Buchanan II, "Government Managers, Business Executives, and Organizational Commitment," *Public Administration Review*, Vol. 34, No. 4, July–August 1974, pp. 339–347; Pfeffer and Veiga, 1999.

[58] Fernandez and Rainey, 2006.

[59] Fernandez and Rainey, 2006.

[60] Buchanan, 1974; Pfeffer and Veiga, 1999.

brought into the organization, the training that the organization provides, and the organization's goals and vision.[61]

Another characteristic of successful organizations is attention to information and analysis that supports continuous process assessment, feedback, and improvement. This should include a rigorous system of data collection and performance evaluation of products, employees, leaders, and processes. This type of assessment can be time-consuming, but the ability to benchmark performance and constantly work to improve it can help keep the organization on track, even as responsibilities, organizational structure, or leadership changes.[62]

Best Practices Suggested by the Literature

The preceding three sections described many factors that could affect the chances of successfully integrating AC and RC staff organizations. There is much overlap in the three bodies of literature, which is not unexpected.

In this section, we present a merged and refined list of these factors. To create this list, the research team started with the full list of factors discovered during the literature search. We narrowed this list by excluding factors that we judged were not applicable or relevant to integrating AC and RC organizations. For example, the practice of providing quick promotions across the boundaries of merged organizations would not be relevant in such cases because AC officers would not be promoted into the RC or vice versa. Some factors associated with successful commercial organizations—such as increasing market share relative to competitors—would also not be applicable to defense organizations in this context.

Next, we identified factors that were very similar or overlapped significantly (for example, "perform organizational self-assessment,"

[61] Fernandez and Rainey, 2006.

[62] Rangone, 1997; Mary Bryna Sanger, "Does Measuring Performance Lead to Better Performance?" *Journal of Policy Analysis and Management*, Vol. 32, No. 1, Winter 2013, pp. 185–203.

"perform organizational analysis," "develop integration measures of effectiveness," and "apply lessons early to prevent problems") and factors that were subsumed in other factors (for example, "remember the importance of people" seems to subsume "treat people fairly" and "promote based on merit"). We combined these factors to create a set of "best practices" suggested by the literature. These practices can serve as a guide when planning an organizational integration and as a framework for assessing progress and making adjustments during and after an integration effort.

Establish the Need and the Vision for Change

Clarifying why an organization needs to integrate or change and defining the future end state and goals of the organization—the "vision"—will be key to successfully implementing changes. If leadership has not clearly articulated the need for change, stakeholders will continue to question whether any change is necessary. At the same time, members of the organization and other stakeholders will want to know what the outcome of the change is going to look like.

Create a Coalition to Support the Change

Successfully implementing proposed organizational changes will rely on the support of both internal and external stakeholders. Lessons from the literature indicate that major organizational change can rarely succeed without leadership support and commitment. If top leadership and managers—both those within the organization and those to whom the organization reports—do not reinforce the changes with their continued support and commitment, implementation can stall and resistance can arise. Leadership at all levels can also set the tone for the integration and ensure that changes do not negatively affect such issues as morale. External stakeholders, such as customers or others who use the organization's products, should also be part of the supporting coalition.

Almost always, some stakeholders are skeptical or outright opposed to any significant changes in organizations. It is important that these voices of opposition be heard during the planning process and that their concerns be considered in the decisionmaking process. If opponents do not feel that their concerns were considered, they might continue to resist implementation of any changes.

Communicate the Vision

In addition to developing a vision for organizational change, leaders need to communicate that vision to others within and outside the organization. The literature notes that change is a process rather than an event. So although communicating the vision for change is certainly part of building a support coalition as described above, the communication should continue throughout the process.

Develop an Implementation Strategy, Including Goals and Measures

The literature indicates that developing and articulating an implementation strategy is a key element of successful organizational change. Well-designed implementation strategies that clarify the goals of organizational change, assign responsibility and accountability, identify risks, and outline mitigation strategies are particularly effective in streamlining the implementation process and can help maintain momentum.

A related point from the literature is that an internal "change agent" or an integration team can be identified to undertake the important task of managing the transition process. Because some changes can take quite a while to effect, it is important to include short-term, as well as long-term, goals in an implementation strategy and to include measures in the strategy that can be used to assess progress toward goals.

Link the Vision and the Structure

The structure of an integrated organization should be linked and consistent with the vision for the organization, rather than being constrained by the structures of any of the previous organizations being integrated. This includes not only the management reporting structure of the organization but also the functions that the organization is undertaking and the people and groups represented in it.

Embed the Changes in the New Culture

When integrating existing organizations, the literature suggests that it is important to embed the changes resulting from the integration in the culture of the new organization. Institutionalizing these changes in the culture, as well as in the organization's policies and procedures, will help to ensure that they endure.

The importance that a healthy organizational culture has for an organization's success is another factor emphasized in the literature. Some of the elements of a healthy culture include flexibility, adaptability, and a focus on the quality of the organization's products. Incorporating these traits in a new organization's culture can help to increase the chances of it succeeding.

Manage the Integration of Cultures

Combining the cultures of integrating organizations can be challenging, and the literature suggests that doing so requires attention and intentional management. One approach is to create an integration team, as mentioned above, to focus on managing the integration of cultures and other aspects of the integration. Ways to integrate organizational cultures include eliminating the idea of protecting the equities of the former organizations in the new one; ensuring representation from the former organizations in all parts of the new organization; and, if applicable, promoting across the former organizations to help eliminate the sense that the former organizations still exist separately.

Maintain Momentum

When organizations integrate or implement other types of changes, it is important to maintain the momentum of change beyond the early phases. The literature suggests that providing the necessary resources and formally institutionalizing any changes through changes in policy and procedures can help to sustain momentum by reinforcing leadership commitment to the changes and signaling to the organization that change remains a priority. These steps can also help to maintain the changes beyond the time when there are changes in leadership.

Remember the Importance of People

People are obviously crucial to any organization's success. Another point emphasized in the literature is their importance when integrating or changing an organization, and resistant people can sabotage even the best-intentioned change. Some of the practices described above, such as building a coalition and communicating the vision, suggest ways to engage people and get their support for change. Other practices mentioned in the literature include selecting for positions and promoting based on merit; empowering people to clear obstacles to change; and treating people fairly, particularly in situations in which jobs are being eliminated. In the last case, it is worth noting that, when people are not treated fairly in layoff situations, the morale of the remaining workers and their commitment to the organization and the changes can also be adversely affected.

Assess Progress and Adjust Accordingly

An organization's willingness to assess its progress is another contributor to success noted in the literature. Determining appropriate measures of effectiveness, collecting the data needed to evaluate the measures, and adjusting direction based on that evaluation are important practices for all organizations, not just those undertaking integration or change. Closely related practices of importance include addressing the source of discovered problems, not just the symptoms, and applying lessons early to help prevent future problems rather than waiting to make changes until problems have already occurred.

The Challenge of Measuring Progress and Success

The practices described above do not easily lend themselves to strictly quantitative measures of success. Many are complicated concepts that are qualitative in nature. It can therefore be challenging to find ways to measure how successfully an organization has been in following them. This can be especially difficult for nonprofit and government organizations that generally cannot use end-result measures, such as profit and market share.[63]

Table 3.1 shows some possible measures associated with each of the practices described above. We do not mean to imply that the measures listed for each practice, when combined, in any way present a

Table 3.1
Possible Measures for Integration Best Practices

Practice	Possible Measure
Establish the need and the vision for change	• A clear statement of the need for change • A clear statement of the vision for change
Create a coalition to support the change	• A list of internal and external stakeholders • Stakeholders who support the change
Communicate the vision	• Number of communications (e.g., briefings, discussion, articles) about the vision to each stakeholder or group of stakeholders
Develop an implementation strategy, including goals and measures	• An implementation strategy, including goals and measures • Responsibility and accountability for each goal and measure • Identified risks and associated mitigation strategies
Link the vision and the structure	• A comparison of alternative structures, including the adopted structure, and their relative consistency with the vision • Consistency of organization's functions and priorities with the vision • Consistency of products with vision • Representation of integrated groups within the organizational structure
Embed the changes in the new culture	• Speed of needed changes in policies and procedures • A statement of quality policy • A quality assessment process • Consistency of organization's processes with quality policy • Assessment of product quality

[63] The literature includes such measures as ways to assess the success of commercial organizations. We did not include them when summarizing our literature review because they generally do not apply to military and other government organizations.

Table 3.1—Continued

Practice	Possible Measure
Manage the integration of cultures	• An integration team • Representation of integrated groups within the organizational structure • Representation of integrated groups at every level of the organization
Maintain momentum	• Speed of needed changes in policies and procedures • Resourcing of implementation strategy
Remember the importance of people	• New-hire training in appropriate methods • Education and training in new methods • Employee morale • Employee retention • Ability to recruit staff with the right talents • Consistency of incentives with vision • Timeliness of product delivery • Customer satisfaction
Assess progress and adjust accordingly	• Explicit policy of assessing measures and feedback • Efficiency (e.g., meeting deadlines) • Innovation (e.g., new business processes) • Changes made as a result of measure tracking • Assessment of measures in implementation strategy

NOTE: Where applicable, we duplicate some measures.

complete assessment of whether an organization has successfully carried out that practice. However, the measures might be useful in indicating where there could be problems, and the changes in some measures might show positive trends or problem areas that need attention.

The best practices discussed in this chapter present a useful framework for examining past integration efforts. The next five chapters present case studies of such efforts undertaken by the five military services—the Air Force, the Army, the Coast Guard, the Marine Corps, and the Navy. Each of these chapters begins with a description of the evolution of AC/RC integration in that service. It then turns to an analysis of current service models of integrated organizational structures and processes. The chapter then identifies lessons learned and concludes by presenting an overall assessment of AC/RC integration efforts in the context of the practices identified above.

CHAPTER FOUR

Active and Reserve Component Integration in the Air Force

The U.S. Air Force has experimented with various forms of AC/RC integration since the 1960s. Over the years, the Air Force has also built strong leadership commitment to the concept of increased integration across the Regular Air Force (RegAF), the Air Force Reserve (AFR), and the Air National Guard (ANG).[1] For instance, in 2016, when discussing the Air Force's efforts to expand total force integration, Air Force Chief of Staff Gen Mark A. Welsh III stated, "We are one Air Force. We are committed to this idea and it's foundational to the way we present our capabilities. We're not going to be successful any other way."[2] The prolonged period of conflict after the terrorist attacks of September 11, 2001, also required more integration across the Air Force's AC and RC. In his 2013 testimony before the NCSAF (a congressionally mandated commission established to examine whether to restructure the Air Force and, if so, how), GEN Charles H. Jacoby Jr., commander of U.S. Northern Command and North American Aerospace Defense Command, reinforced this by saying, "in the last 10 years we have gotten closer to the Total Force than ever because we shed blood together, so let's not undo that. How we do it is as impor-

[1] In this study, we did not include the Civil Air Patrol, the civilian auxiliary of the Air Force, which operates as a nonprofit corporation. It consists of volunteers who provide emergency services, aerospace education, and cadet programs (Civil Air Patrol, "Online Media Kit," undated).

[2] U.S. Air Force, "Air Force Continues to Pursue Total Force Integration," press release, March 11, 2016.

tant as what we do."[3] However, despite this progress, there still remain challenges in integrating the Air Force AC and RC.

Evolution of Active and Reserve Component Integration in the Air Force

The Air Force has a long history of associating active and reserve units, going back to the Vietnam War era and predating DoD's adoption of the Total Force policy by several years. The first Air Force "associate units" (units in which AC, AFR, and ANG units combined) were established in 1968, when the 63rd Military Airlift Wing (an AC unit) assumed operational control of the 944th Military Airlift Group (an AFR unit) for unit training assemblies, active-duty training periods, and any aircrew members when they integrated into the active wing. Within days of the first training assembly, a reservist served on an operational mission, and, a few months later, an all-reservist C-141 crew from the 944th flew a mission in Southeast Asia for the first time. By 1974, four C-5 and 13 C-141 reserve squadrons were aligned under Air Mobility Command wings at six installations.[4]

This same approach continued for about 30 years, with RC units being subordinate to AC units. But in the early 2000s, the Air Force began experimenting with new forms of associated units in which two units from different components would operate as an integrated unit for mission purposes, but each component would have its own detachments to provide administrative control and support to the personnel from their respective components. Over the years, these detachments evolved to full-blown, separate organizational structures and chains of

---

[3] GEN Charles H. Jacoby Jr., commander, U.S. Northern Command and North American Aerospace Defense Command, testimony before the National Commission on the Structure of the Air Force, September 26, 2013.

[4] Gerald T. Cantwell, *Citizen Airmen: A History of the Air Force Reserve, 1946–1994*, Air Force History and Museums Program, 1997, pp. 311–312.

command for the units from each component. The three types of associations are as follows:

- **classic association:** an integration model that combines active and reserve elements, with the AC retaining principal responsibility for a weapon system and sharing the equipment with one or more RC units[5]
- **active association:** an integration model that combines active and reserve elements, with the RC retaining principal responsibility for a weapon system and sharing the equipment with one or more AC units[6]
- **air reserve component association:** An integration model that combines two RC elements, with one retaining principal responsibility for a weapon system and sharing the equipment with one or more of the other component's units.[7]

Because the Air Force Reserve Command (AFRC) was established on February 17, 1997, a more formal, dual command structure evolved in these associations than was present at their inception.

National Commission on the Structure of the Air Force

The NCSAF was created in the National Defense Authorization Act for Fiscal Year 2013 to

> undertake a comprehensive study of the structure of the Air Force to determine whether, and how, the structure should be modified to best fulfill current and anticipated mission requirements for the Air Force in a manner consistent with available resources.[8]

---

[5] NCSAF, 2014, p. 29.

[6] NCSAF, 2014, p. 29.

[7] NCSAF, 2014, p. 29.

[8] NCSAF, 2014, p. 7; Public Law 112-239, National Defense Authorization Act for Fiscal Year 2013, January 2, 2013.

The NCSAF made 42 recommendations for restructuring the Air Force. One of those recommendations was to eliminate the redundant nature of the dual command structure that evolved in the associations described above. Both for fiscal reasons and as a means to a more effective integrated structure, the NCSAF instead proposed the establishment of "integrated wings" ("i-Wings"). In its report, the NCSAF recommended that, "in the i-Wing, unit leadership positions, both officer and enlisted, be filled by personnel of both components that make up the associate unit, *and the unit operate as a single entity rather than two, side-by-side commands*" and noted further that "the i-Wing is a logical extension of the forward-thinking approach first instituted by associate units."[9]

But the NCSAF went even further in its recommendations for achieving total force integration. If the Air Force were to adopt increased integration through the "logical extension of operating as a single entity rather than two side-by-side commands," the commission reasoned,

> the need for an Air Force Reserve Command as a "force providing" headquarters declines, as does the need for its subordinate Numbered Air Forces. Commanders of operational major commands (Air Combat, Mobility, Space, etc.) and their Numbered Air Forces can make decisions regarding the employment of integrated Air Force capabilities. The Commission believes the current mission of the Air Force Reserve Command and its Numbered Air Forces can be disestablished.[10]

This was the only one of the NCSAF's 42 recommendations that Air Force leadership summarily rejected. Although the NCSAF did conclude that, after a long-term evolutionary process, the role of the AFRC would become redundant, the commission affirmed

> the retention of the positions of the Chief, Air Force Reserve and the Director, Air National Guard with direct access to the Chief

[9] NCSAF, 2014, p. 29. Emphasis ours.

[10] NCSAF, 2014, p. 31.

> of Staff and with small, but sufficient staffs to properly advise Air Force leadership on policies necessary to recruit, retain and sustain talented and motivated Airmen in both the Air Force Reserve and Air National Guard.[11]

Current Air Force Models of Integrated Organizational Structures and Processes

The Air Force is in the process of integrating at both the staff and unit levels. We discuss each of these below.

Total Force Integration at the Air Force Staff Level

Prior to the NCSAF standing up in June 2013, the Air Force stood up the Total Force Task Force (TF2) in late January 2013. TF2 transitioned to the Total Force Continuum (TFC) office on October 1, 2013. The mission of TF2/TFC office was to determine how best to integrate the Air Force's three organizations into an effective total force. One of the recommendations that the TFC office put forth was to do a better job of component integration at the staff levels, including Headquarters, Air Force (HAF); chief, AFR staff; and director, ANG staff. The NCSAF affirmed that goal and included it as one of its recommendations.[12]

The Air Force's goal is to integrate staffs at multiple levels, but it began with the HAF. Because integration deals with personnel, the Office of the Deputy Chief of Staff of the Air Force for Manpower, Personnel and Services (A1) stood up to integrate first and began this effort in 2014. As background, between 2010 and 2012, there was an effort to merge the A1 (which is part of the Air Staff and reports to the Chief of Staff of the Air Force) with Office of the Assistant Secretary of the Air Force for Manpower and Reserve Affairs (which is part of the Air Force Secretariat and reports to the Secretary of the Air Force) but, after pushback from Congress, went back to the status quo. The

[11] NCSAF, 2014, p. 31.

[12] NCSAF, 2014, p. 33.

current A1 integration effort brings in 26 Active Guard Reserve (AGR) personnel from the AFR and the ANG, representing about 10 percent of the A1 staff. The understanding, fully coordinated with the respective components, is that the RC AGRs will bring their respective RC portfolios with them. That said, the expectation is that, with the passing of time and working together, an airman from any component will be able to process actions for any of the other components.

After the A1, it is yet to be determined which staff directorates will integrate next. One Air Force interviewee indicated that some directorates and their associated tasks, such as the A1 and the Office of the Deputy Chief of Staff of the Air Force for Logistics, Engineering and Force Protection are more conducive to integration than others, such as the Office of the Deputy Chief of Staff of the Air Force for Operations. Following are the principles for A1 staff integration success identified during our discussions with Air Force leadership:

- Define integration and its goals up front in terms of total force and total force staff.
- Define expectations up front.
- Tailor the numbers and types of positions that RC airmen will fill to what makes sense. Design for success.
- Ensure that multiple key positions are assigned to RC airmen.
- Ensure that resources for the components are planned together.

In addition, our discussions identified challenges with this integration effort, including that the AFR and the ANG do not have deep benches from which to draw to select talented airmen to serve on the HAF.[13] They also indicated that integration works best with Title 10 (AC and AFR) because of the extra layers of exceptions to policy, memoranda of understanding (MOUs), and memoranda of agreement (MOAs) that are required because of Title 32.

---

[13] This is not to imply a qualitative difference between airmen in the components but is a reflection of the numbers available from which to draw because of the much smaller numbers of full-time RC airmen.

Total Force Integration at the Unit Level

An important piece of total force integration at the unit level is integration of leaders among the components. For instance, the Air Force plans to fill key leadership positions with cross-component airmen. Currently, three AFR officers are set to command AC units, including two maintenance squadrons and a fighter wing. Those reserve officers will parallel the four AC officers who are currently serving in wing or vice wing command positions in both ANG and AFR units. Similarly, the Air Force Chiefs Group actively considers chief master sergeants from both the AC and the RC for certain senior enlisted billets.[14]

The Air Force also developed more-detailed guidance on execution of unit associations. Air Combat Command (ACC) issued a supplement to AFI 90-1001, *Responsibilities for Total Force Integration*, in 2010—three years after it was first published. In 2014, the Office of the Deputy Chief of Staff of the Air Force for Strategic Plans and Requirements issued a detailed guidance memorandum to accompany AFI 90-1001 that included templates and worksheets for preliminary documentation of the associated units. Of particular importance are (1) operational direction (OPDIR) to enable day-to-day direction of a multicomponent workforce and (2) automatically executing Title 10 orders (in place before AFI 90-1001 was published). We examine each of these below.

Operational Direction

Unity of command and unity of effort are difficult in associations because of both statutory and organizational structure. Using the example of a classic association, each component's element has its own, separate chain of command that executes administrative control over its airmen. By statute, total integration of command structures or organizations cannot appear within a single unit manning document. For example, Title 10 commanders cannot direct Title 32 airmen associated with them, and, even when the Title 32 airmen perform operational missions in a Title 10 status, the Title 10 commander does not execute Uniform Code of Military Justice (UCMJ) authority over

---

[14] U.S. Air Force, 2016.

them: They return to Title 32 status and state disciplinary control at the conclusion of the operational mission. To help mitigate this situation, the Air Force established the concept of OPDIR. OPDIR is not a joint doctrinal term and applies in the Air Force only to associations.

According to Air Force Guidance Memorandum to AFI 90-1001, 2014:

> Operational Direction is defined as "the authority to designate objectives, assign tasks, and provide the direction necessary to accomplish the mission or operation and ensure unity of effort." Authority for operational direction of one component member over members of another component is obtained by agreements between component unit commanders (most often between Title 10 and Title 32 commanders) whereby these component commanders, in an associate organizational structure, issue orders to their subordinates to follow the operational direction of specified/designated senior members of the other component for the purpose of accomplishing their associated mission.[15]

OPDIR, as defined above, is not a command authority because that is prohibited by law. But it is an authority to "designate objectives, assign tasks, and provide necessary direction to achieve unity of effort." But in order for that to work, because of the law, well-written and agreed-to MOUs and MOAs must be developed that, among other things, call for the Title 32 commander to give orders to his or her subordinate Title 32 airmen to follow the OPDIR provided by the Title 10 host commander. OPDIR, then, is about as close as the Air Force can get to unity of command and unity of effort without changing existing laws and policies. Figure 4.1 illustrates both the complexity of the OPDIR relationship and what the NCSAF was getting to in its recommendation to ultimately eliminate redundant and costly dual chains of command.

According to AFI 90-1001, the objective of the total force integration program is "to meet Air Force operational mission requirements by aligning equipment, missions, infrastructure, and manpower

---

[15] Moeller, 2014, p. 61.

Figure 4.1
U.S. Air Force Notional Classic Association Illustrating Operational Direction

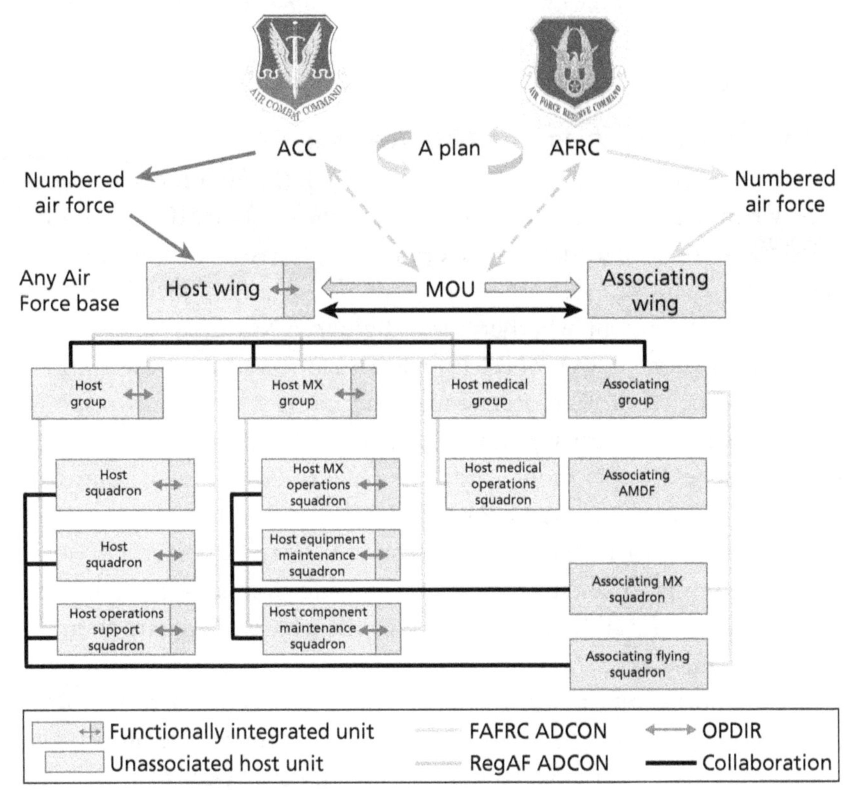

SOURCE: A-Plan Org Chart.pptx, briefing slide, undated.
NOTE: MX = maintenance. AMDF = aero medical dental flight. ADCON = administrative control. This organizational chart is for a notional classic association for an ACC host fighter unit and an AFRC fighter unit. Substitutions must be made for other types of units, associations, and components (e.g., the ANG does not have a numbered air force).

resources within the Air Force to enable *a more effective and efficient use of these assets*."[16] The i-Wing is an effort to move further down the path to total force integration by incorporating in its initial pilot program at Seymour Johnson Air Force Base a dual-hatted construct to minimize the necessity for two separate chains of command.

---

[16] AFI 90-1001, p. 4. Emphasis ours.

Current statutes still mandate the creation of artificial constructs simply to fulfill the requirements of the law (even this pilot program required 25 exceptions to policy),[17] but it is an evolutionary step that at least eliminates having two separate chains of command at the group level through the creation of dual-hatted positions.

Automatically Executing Title 10 Orders

The construct of automatically executing Title 10 orders is necessary when an ANG unit or element performs Title 10 operational missions—some of which occur on a routine basis, such as the 24/7 Aerospace Control Alert mission set.[18] These automatically executing orders are present whether a stand-alone ANG unit performs the Title 10 mission or the ANG unit is an associate of an AC unit in a classic association. The orders are a mechanism created in response to two separate statutory requirements. One is that any elements performing Title 10 operational missions have to be under an unbroken Title 10 chain of command for the duration of the execution of the mission. The other is that Title 32 airmen have to be under a Title 32 chain of command when not performing Title 10 missions and have to be available for recall by the governor of their particular state should a state emergency arise.

A standing, automatically executing order specifies a clearly identifiable factual event, the occurrence of which activates the orders, and another clearly identifiable factual event, the occurrence of which terminates the order.[19] One example of these types of orders is the aforementioned Aerospace Control Alert mission, which is a Title 10 mission performed predominantly by Title 32 ANG units. The pilots are in Title 32 status while on the ground, but, as soon as the wheels of the aircraft leave the runway, the automatically executing order is activated, and the pilot transitions to Title 10 status until the wheels of

[17] Discussion with senior officials from the TFC office, May 13, 2016.

[18] The Aerospace Control Alert mission is "a national network of fully loaded aircraft ready to protect the country on a moment's notice . . . to intercept, inspect, influence, and if necessary, defeat a potential airborne threat" (Colorado National Guard, "Aerospace Control Alert Mission," November 2015).

[19] AFI 90-1001, p. 65.

the aircraft hit the runway upon landing, and the pilot is once again in Title 32 status.

In almost a summary statement rolling up the necessity for total force integration plans, and including the need for such mechanisms as OPDIR and automatically executing orders, AFI 90-1001 contains the following:

> These integration plans must address command relationships between federal and state authorities, and the determination of appropriate duty statuses when performing the federal mission. This presents challenges to Air National Guard units in that they must ensure the unit stand-up and personnel duty statuses *are in compliance with the current statutory framework*, and that personnel performing a Title 10 mission are attached to the appropriate Combatant Commander. *This is especially necessary because the current statutory framework has not caught up with the Air Force and Air National Guard leadership's goal of "Total Force Integration."*[20]

This can be interpreted as saying that the manner in which senior Air Force leadership desires to operate in a truly integrated way is out ahead of statutes that never envisioned this kind of integration for ongoing operational missions.

Lessons Learned

Given the friction among the components, particularly between the RegAF and the ANG, and as a result of conclusions reached by the NCSAF and the TF2/TFC, the Air Force realized that it needed a new vision for meeting Air Force needs in a reduced-budget environment. Specifically, the Air Force studied and determined the optimal AC/RC force mix across all of its mission areas to "take care of people, balance today's readiness with tomorrow's modernization, and making every dollar count."[21] Broadly, the Air Force realized that it could not

[20] AFI 90-1001, p. 62. Emphasis ours.

[21] U.S. Air Force, 2016.

achieve these goals without better efforts directed at AC/RC integration, although the forms and degree of integration are still being determined. So although there is a broad vision for increased integration, there is not yet a well-defined end state.

During its integration efforts, the Air Force learned that detailed agreements were needed to effectively integrate forces operating under Title 10 and Title 32. But it also learned that sufficiently detailed and creative agreements could address most of the statutory challenges involved and could achieve a workable level of unity of command.

Assessment of Active and Reserve Component Integration in the Air Force

The Air Force has a history of integration, but, in the wake of the NCSAF report, it is involved in determining and pursuing a new level of integration. The Air Force has been using the recommendations from both its own TFC office and the NCSAF as a guide to implementing total force integration. After beginning successful integration efforts in the late 1960s, the Air Force has continued to experiment and to seek the best models for total force integration at both the unit and staff levels. In order to be more successful, the Air Force needs to better define its idea of what truly constitutes a total force and what limits it is unwilling to exceed. Currently, the Air Force is following a set of principles for total force integration that includes each of the components retaining ADCON, being responsible for that component's program authorization, and providing its own professional development. Just as in the cases of the other services, the Air Force will have to decide how integrated it wants its total force to be or whether it will be content with individual components competing for their own equities while operating in an integrated manner to the maximum extent possible without truly integrating.

The Air Force's integration efforts reinforce some of the best practices and success factors identified in Chapter Three. For instance, by all accounts, the coalition among the three Air Force organizations is the most congenial it has been for a long time; however, this might be because integration is not being pushed to a degree that the organizations find threatening. The Air Force is methodically moving forward developing plans and testing various forms of organizational structures, complete with goals and measures, but it has not yet developed a clear vision for what its total force will look like and how to implement it.

Without changes to law and policy, the structural integration of Title 10 and Title 32 forces can go only so far, regardless of the potential perceived benefits of going further. As noted in this chapter, without changes, integration efforts involving the ANG are forced to incorporate many work-arounds, such as multiple dual-hats for commanders, exceptions to policy and a lack of true unity of command.

In terms of other best practices identified in the literature, although the Air Force is maintaining the momentum of the process and is looking for innovative ways to take care of its people, the cultural and structural vision for cross-component integration is not yet defined enough to establish how to embed it, maintain it, and assess it.

The Air Force's integration efforts also reveal some new best practices not identified in the literature (e.g., mechanisms for unity of effort). For instance, the Air Force's use of OPDIR is a novel mechanism to achieve cross-component unity of effort. The Air Force's integration efforts also reinforce some barriers to integration identified in Chapter Three (e.g., lack of a clear vision, statutory and policy barriers).

CHAPTER FIVE

Active and Reserve Component Integration in the Army

Like in the Air Force example, the Army has, for decades, experimented with various approaches to improve the integration between its AC and RC. Prior to 9/11, the RC served as a strategic reserve that would be called to duty only in the event of a major conflict. During the operations in Afghanistan and Iraq, RC forces were deployed regularly as an operational reserve. As those operations wound down, the demand for deployed forces and the supplemental funding for wartime operations both decreased, which resulted in calls to reduce force structure and budgets. This drove the Army (along with the other services) to try to identify the optimal AC/RC force mix while maximizing readiness and maintaining operational use of the RC so as not to lose the proficiencies and trust gained over the previous decade.

Evolution of Active and Reserve Component Integration in the Army

As mentioned in Chapter One, in August 1970, Secretary of Defense Laird directed the services to incorporate a total force concept into all aspects of planning, programming, manning, equipping, and employing reserve forces.[1] At the time, this directive coincided with the pending abolition of the draft as a source to augment regular Army forces, cuts to defense spending, and preserving as much capacity as possible

[1] Melvin R. Laird, Secretary of Defense, "Support for Guard and Reserve Forces," memorandum, Washington, D.C., August 21, 1970, as quoted in Rostker et al., 1992, p. 33.

to deter threats. What followed was more than 45 years of experimentation by the Army to determine the best means of implementing a total force.

The Army continues to talk a lot about the total force, total force integration, and making the best use of its three organizations—the active Army, the U.S. Army Reserve (USAR), and the Army National Guard (ARNG). The most recent policy in that regard came from the Secretary of the Army in September 2012.[2] As highlighted by several quotations below, total force policy pronouncements start with assertions that it is a good thing, and that the actions prescribed will make the total force a more effective respondent to national security needs. But the Army has failed to articulate: what a total force is and how it operates as a total force rather than as three entities competing for finite resources, what specific problems it is trying to solve with increased integration, what models of integration for which types of approaches will solve those problems, and clear metrics and measures of success. Consider the following extracts from a 2000 GAO report. They still pertain today:

> For nearly three decades, the Department of Defense has had a "total force" policy in place aimed at maintaining the smallest possible active duty force and complementing it with reserve forces. As the military downsized in the 1990s, it increased its emphasis on the total force concept and sought new ways to use both active and reserve components effectively. The Department of Defense has emphasized the importance of integration as one way to do this, but without clearly defining integration.
>
> In 1999, the Army Chief of Staff said that completing the full integration of the active and reserve components was one of his six main objectives. However, like the Department of Defense, the Army has yet to define what it means by full integration.
>
> For example, in 1997 the Secretary of Defense issued a two-page memorandum that called for "a seamless total force" and the

---

[2] McHugh, 2012.

> elimination of all residual barriers to effective integration. While the memorandum included four basic principles of integration, such as "leadership by senior commanders—Active, Guard, and Reserve—to ensure the readiness of the total force," it did not contain measurable results-oriented goals to evaluate the services' integration progress.[3]

Current Army Models of Integrated Organizational Structures and Processes

Integration Efforts on the Army Staff

Formal efforts to incorporate RC personnel, both military and civilian, began with a review in November 2000 that looked at Headquarters, Department of the Army (HQDA) transformation from four aspects. The first was culture and examined the differences in roles between uniformed and civilian personnel. The second looked at process and ways to streamline the bureaucracy. The third focused on the structure of HQDA, and the fourth looked at improved use of information technology and automated systems to increase efficiency. As part of the third effort, then–Secretary of the Army Thomas E. White created the Realignment Task Force (RTF), charged with eliminating overlap and redundancy in functions and layers of supervision.[4] One example of the results of RTF recommendations was the downsizing of the Office of the Chief of the Army Reserve (OCAR) and the integration of some of those personnel into various directorates across the Army staff. A joint memorandum signed by the Chief of Staff of the Army and the Secretary of the Army on January 4, 2002, directed the implementation.[5] In his memorandum acknowledging compliance, then–Chief of the Army Reserve (CAR) LTG Thomas J. Plewes agreed to implement

---

[3] GAO, *Force Structure: Army Is Integrating Active and Reserve Combat Forces, but Challenges Remain*, Washington, D.C., GAO/NSIAD-00-162, July 2000, p. 5.

[4] Christopher N. Koontz, *Department of the Army Historical Summary: Fiscal Year 2001*, Washington, D.C.: U.S. Army Center of Military History Publication 101-32-1, last updated November 17, 2011, pp. 4–5.

[5] Secretary of the Army and Chief of Staff of the Army, 2002.

a cut from 279 authorizations to 120 authorizations, with a temporary increase of 20 AGRs during the transition period. The understanding was that the previous OCAR staff members transitioning to the Army staff would take their functions with them and that the success of the realignment would depend on the Army Staff's acceptance of those functions.[6] The staff members who moved were spread out among many offices in HQDA.

But the change did not go well. As opposed to the expectations being set in the Air Force A1 case, the AC personnel on the Army staff perceived these AGRs from the USAR as a means to fill their own manning shortfalls created by other RTF decisions. The AC personnel assigned other duties to the AGRs. Consequently, a large portion of the functions and actions that these AGRs performed when they were assigned to the OCAR staff were not getting done. This put not only the remaining OCAR staff but also the USAR overall at a disadvantage. One civilian staff member told us that, because of their own manning shortfalls and the functions not being performed by the reassigned AGRs, "we lost battles we didn't even know we had a chance to fight." Members of the OCAR staff with whom we spoke had several other summary comments:

- Roles and functions that were supposed to transfer with the AGRs and continue to be performed were not.
- None of the agreements was codified into regulation (this was even more important given that the total Army and the Army Staff were focused on supporting operations in Iraq and Afghanistan right after this realignment).
- USAR equities were not represented commensurate with the numbers of AGRs transferred.
- In order to make up the perceived shortfalls, the USAR had to beef up the U.S. Army Reserve Command to operate in conjunction

[6] CAR, 2001.

with OCAR as "one staff/two locations" in order to adequately follow and represent USAR equities.[7]

Corps and Division Headquarters Pilot

Although the motivation for pursuing a multicomponent unit (MCU) or a multicomponent approach is often to increase training readiness of the RC or to foster better implementation of Army Total Force Policy, the establishment of a pilot study for converting corps and division headquarters into integrated, multicomponent staffs was driven by yet another reason: to mitigate the effects of AC downsizing.[8] The units chosen for the pilot study were the XVIII Airborne Corps Headquarters at Fort Bragg, North Carolina, and the 101st Airborne Division at Fort Campbell, Kentucky.

XVIII Airborne Corps

Under the integrated multicomponent staff pilot construct, the XVIII Airborne Corps carries all unit equipment and personnel under two derivative unit identification codes (DUICs) organic to the corps' headquarters and headquarters battalion (HHBn). One is active Army with 619 soldiers authorized and required. The second is USAR, with 56 soldiers authorized and required, for a total of 675 spaces.[9] The CAR agreed to fund five additional full-time staff to be assigned to the corps HHBn to facilitate support actions required by the USAR DUIC. In addition to all of the personnel and equipment being on the same document for the corps headquarters, this unit has been set up to be a fully integrated total force unit to the maximum extent current statutes will allow.

---

[7] We base the preceding paragraphs on information gathered during a discussion with members of the OCAR staff on February 11, 2016.

[8] Chris Reddish, Forces Command, and Mark Berglund, National Guard Bureau, briefing on multicomponent units to the National Commission on the Future of the Army, Washington, D.C., August 17, 2015.

[9] Commander, U.S. Army Forces Command (FORSCOM), and CAR, "Implementation of the XVIII Airborne Corps Multiple-Component Unit (MCU) Pilot," memorandum of agreement, May 8, 2015, p. 2.

The XVIII Airborne Corps commander is responsible for the training and readiness of the corps headquarters as a total force MCU. This includes being responsible for the Soldier Readiness Processing for both AC and RC soldiers. In contrast to the authorities allowed AC commanders under the Army associated unit construct (in which the AC commander only approves the associated RC unit's training plan but does not oversee its execution),[10] the XVIII Airborne Corps commander both plans and executes the integrated training plan for the XVIII Airborne Corps Headquarters, taking into account the limitations of the part-time status of the USAR unit members. This practice allows for the maximum amount of effectiveness and flexibility in incorporating and fully using the capabilities provided by the USAR soldiers.

The commanding general, XVIII Airborne Corps, and Fort Bragg, North Carolina, can exercise general court-martial convening authority and original and appellate jurisdiction over USAR soldiers assigned to the corps headquarters staff and performing duty at Fort Bragg. The granting of this authority to the AC commander highlights how fully the Army is trying to integrate this unit.

As stated previously, this corps headquarters integrated staff MCU pilot embodies many of the key points that enhance integrated staff MCU effectiveness:

- multiple components on a single manning document
- unity of command
- UCMJ authority over the entire unit
- single readiness reporting chain
- colocation of the entire unit, both AC and RC elements
- shared equipment for the unit's requirements
- allocation of additional training days.

[10] Secretary of the Army, "Designation of Associated Units in Support of Army Total Force Policy," memorandum for principal officials of Headquarters, Department of the Army, and commanders, U.S. Army Forces Command, U.S. Army Training and Doctrine Command, and U.S. Army Pacific, March 21, 2016, ¶ 3.

Although this model still requires certain processes to be done through USAR channels, those processes are executed under the direction of the AC chain of command by full-time support (FTS) personnel assigned to the corps headquarters.

101st Airborne Division

Under the integrated staff MCU pilot construct, the division carries all unit equipment and personnel under five DUICs organic to the division HHBn. One is active Army, with 481 soldiers authorized and required. Two are ARNG. One of these is from the Wisconsin ARNG, with 66 soldiers authorized and required. The second is from the Utah ARNG, with 53 soldiers authorized and required. There are also two FTS DUICs to provide support (one from the USAR and one from the ARNG, with five and four soldiers authorized and required, respectively), for a total of 609 spaces (481 AC and 128 RC).[11]

This unit is also set up to be an integrated total force unit to the maximum extent possible, although geographic separation and Title 32 constraints introduce challenges not present in the pure Title 10 XVIII Airborne Corps model. In addition to the geographic separation, the challenges come directly from Title 32 concerns and the extra layers of coordination and shared ADCON that stem from a Title 10/Title 32 mix. In the sections that follow, we highlight those differences, as described in the MOA.

In the corps model, FORSCOM's role was primarily to certify the actions of the XVIII Airborne Corps commander. In the 101st Airborne Division model, FORSCOM gains a coordination role between the division, the director of the ARNG, and the commander of U.S. Army Reserve Command.[12] In the XVIII Airborne Corps model, the corps commander plans, approves, and executes the MCU's training plan. A major difference in this division model is that the division commander "directs" the training plans and priorities for the Title 10 elements of the unit but is limited to "providing guidance and approving the training plans" for the Title 32 ARNG soldiers assigned to the

---

[11] Commander, FORSCOM, et al., 2016.

[12] Commander, FORSCOM, et al., 2016, ¶ 5.a.

unit.[13] This is primarily because there is not unity of command in the MCU mixing Title 10 and Title 32 soldiers. As specified clearly in the MOA,

> At all times until mobilized, Title 32 Soldiers will be under the command and control of a Title 32 Commander. A limited form of tactical control (TACON) for training may be exercised on the part of the Title 10 Commander, in coordination with the Title 32 Commander who retains administrative control (ADCON) and operational control (OPCON) over the Title 32 Soldiers.[14]

The commander, 101st Airborne Division, and Fort Campbell, Kentucky, can exercise general court-martial convening authority and original and appellate jurisdiction over active-duty, USAR, and ARNG personnel in a Title 10 status. The adjutants general retain command and UCMJ authority over the Title 32 soldiers from their states.[15] This is another example in which the level of integration and unity of command is lower in MCUs that mix Title 10 and Title 32 soldiers.

The USAR and ARNG full-time elements are colocated with the AC unit at Fort Campbell, Kentucky. The other ARNG elements are in Wisconsin and Utah. As a result, they perform their normal weekend drills at their home stations and link up with the other elements during annual training.

MG Gary J. Volesky, the commanding general of the 101st Airborne Division (Air Assault) and Fort Campbell, noted several concerns in his initial assessment of the MCU's mission effectiveness.[16] He described the MCU as "partially" closing the gaps created by force reductions (723 AC to 481 AC and 129 RC). He further explained that capacities and capabilities are either eliminated, reduced without

---

[13] Commander, FORSCOM, et al., 2016, ¶ 8.a.

[14] Commander, FORSCOM, et al., 2016, ¶ 8.d.

[15] Commander, FORSCOM, et al., 2016, ¶¶ 9.a, 9.b.

[16] Gary J. Volesky, major general, U.S. Army, commanding general of the 101st Airborne Division (Air Assault) and Fort Campbell, "101st Airborne Division (Air Assault) Multi-Component Unit (MCU) Assessment of Mission Effectiveness," memorandum for commanding general, U.S. Army Forces Command, December 16, 2015.

corresponding RC backfill, or reduced by as much as 50 percent. He also noted that the AC division is left without critical capabilities for training because of both geographical separation and part-time status. Another major concern that General Volesky expressed is the lack of sufficient capacity in the RC DUICs to enable the division headquarters to surge within its Army force generation cycles. Finally, he noted that, if the RC soldiers are limited to the standard 38 or 39 training days, the result would be degraded proficiency for the entire division staff. Although he focused primarily on the number of days, the problem is exacerbated when a large portion of those training days are conducted separately from the rest of the division headquarters team.

Like with the corps headquarters integrated staff MCU, this division headquarters MCU pilot embodies many of the key points that enhance MCU effectiveness: one or more components on a single document, single unit status reporting chain, shared equipment for the unit's requirements, and allocation of additional training days. But this model also lacks aspects that truly bring an MCU together in an integrated way: unity of command; UCMJ authority over the entire unit; colocation of the entire unit, both AC and RC elements; and shared equipment for the unit's requirements. In MG Brian J. McKiernan's assessment, he commented on both the individual soldier model and colocation:

> The individual Soldier model implemented by the XVIII Corps MCU appears to provide a more effective solution. The Non Co-located . . . model implemented by the 101st Division did not provide the same flexibility in integration and involved greater challenges with developing and maintaining training plans while managing individual requirements for training days.[17]

This model also introduces some additional approval layers, such as having to secure the state governor's assent for additional training days and the provision that allows adjutants general to pull back an ARNG

[17] Brian J. McKiernan, major general, U.S. Army, "Multi-Component Unit (MCU) Final Findings and Recommendations," memorandum for commanding general, U.S. Army Forces Command, December 18, 2015, p. 14.

Title 32 member at any time, even if not to fulfill some higher Army priority. Although this model still requires certain processes to be handled through ARNG and USAR channels, those processes are still executed under the direction or request of the AC chain of command by FTS assigned to the corps headquarters.

Although not the solutions to which components would gravitate if funding were not an issue and with only an initial short-term assessment of the division model, both models illustrate the value of the RC filling gaps and shortfalls as a result of budget constraints and AC force reductions. Because the cuts were steep and the RC lacks the available force structure and sufficient numbers of soldiers qualified in the military occupational specialties required, these MCU solutions might not be as effective as their previous all-AC counterparts. But these multicomponent corps and division headquarters clearly have more capability and capacity than they would have had otherwise after forced reductions and absent the addition of the RC DUICs. Although there is much potential in these MCUs as an AC shortfall mitigation strategy, "[f]urther analysis is required to determine the long term sustainability of the MCU program before Army-wide implementation. Funding commitments, sourcing types of headquarters, and MTOE [modified table of organization and equipment] refinement must all be addressed to better support a sustainable model."[18]

Lessons Learned

Among the services, the Army is most constrained at pursuing approaches to integration by size, structure, and culture. It being the largest of the services, size alone makes major change more difficult. Although both the Air Force and the Army were structured to provide a strategic reserve, the Air Force has a much longer history of operational integration, and airmen operating different platforms do so on more of an individual basis than in large, collective formations. In the Army, integration efforts to date are most successful when focused

[18] McKiernan, 2015, p. 14.

more on individuals and teams making up regular Army shortfalls or augmenting regular Army units and staffs. Some of the other key lessons that arise from the Army's integration experiences are the importance of unity of command and colocation.

Assessment of Active and Reserve Component Integration in the Army

The Army has long experimented with different approaches and degrees of integration. Given friction among the components, particularly between the active Army and the ARNG over the Army's Aviation Restructuring Initiative, and as a result of conclusions reached by the NCFA, the Army is looking at new approaches to integration at both the staff and unit levels. Although the Army is pursuing these new approaches, it lacks a vision with a clearly defined end state. Like with the Air Force, there is general acceptance that different force-mix ratios are likely required to balance current readiness with future modernization, but, in terms of what form that should take and what degree of integration is both desirable and achievable, the Army is still searching.

The Army's integration efforts reinforce the need for some of the best practices and success factors identified in Chapter Three. For instance, the extent of the Army's vision today is that integration will likely improve readiness, will better achieve a "one-Army" culture, and will make better use of the different strengths of the components and mitigate AC shortfalls, but a clearly defined end state of what an integrated Army looks like that accomplishes those goals is lacking. In addition, practices associated with developing and establishing plans that stem from the vision are present but disjointed.

As the only other service (other than the Air Force) with a Title 32 organization, the Army faces the same statutory issues that limit the range of integration and unity of command and impose a limit on the range of options that can be considered. Like with the Air Force, this does not mean that integration with the regular Army and the ARNG is not possible—it just means that it is more difficult to achieve.

The bottom line is that the Army will need to decide what level and type of integration it thinks would be beneficial, and then to decide what it thinks is either achievable or worth fighting for in terms of changes in law, policy, and structure to achieve those perceived benefits.

The Army's integration efforts also reveal some new best practices not identified in the literature (e.g., mechanisms for unity of command and colocation). The Army's integration efforts also reinforce some barriers to integration identified in Chapter Three (e.g., lack of a clear vision, statutory, and policy barriers).

CHAPTER SIX

Active and Reserve Component Integration in the Coast Guard

The U.S. Coast Guard has an AC with 36,000 personnel and a federal RC with 7,000 personnel. Prior to 1994, the U.S. Coast Guard Reserve (USCGR) operated separately from the active Coast Guard, although both components shared the same equipment that they do today. Our discussions with Coast Guard leadership indicate that, culturally, the active side held the reserve in some disdain. There was some use of the USCGR to augment operational missions, but there was confusion in the force as to the competing concepts of augmentation and preparing for mobilization.[1]

Evolution of Active and Reserve Component Integration in the Coast Guard

Up until 1994, the AC and RC of the Coast Guard operated on separate pay and personnel systems. In February 1994, the commandant of the U.S. Coast Guard (CG-00) tasked a working group to investigate the potential for active and reserve integration. The catalyst was an effort to generate maximum productivity from the combined resources of the Coast Guard's AC and RC. The working group made its recommendations, and, on August 12, 1994, the commandant directed the

[1] There is also a Coast Guard Auxiliary, which is made up of civilian volunteers who focus primarily on boating safety and are therefore not included in this study (U.S. Coast Guard Auxiliary, "About the Auxiliary," last updated July 14, 2015).

integration of the two components. Many actions occurred in a very short time frame, including these:

- Virtually all reserve units would be colocated with active units.
- The active Coast Guard assumed responsibility for training the USCGR and writing their personnel evaluation reports.
- The USCGR received all of its taskings from an active Coast Guard chain of command.
- The USCGR was integrated into active Coast Guard day-to-day and surge operations as required and as the USCGR was available. The concept that the best training to prepare for mobilization was to conduct operational missions was reiterated.
- An integrated pay and personnel system was created.[2]

A major change such as this, particularly because it was executed quickly, was not without its challenges. The USCGR decreased from 12,000 to 8,000 personnel, losing about one-third of its officer corps.[3] When the active Coast Guard gained responsibility for the USCGR, they became responsible for nearly twice as many coastguardsmen almost overnight. Not only was no additional manpower provided, but the active Coast Guard commanders and staffs also lacked familiarity with reserve-specific issues. To address this problem, the Coast Guard took AGRs and created integrated support commands, which were placed within district commands to provide support to the reservists. In 2009, these additions were expanded by creating the Reserve Force Readiness System, a "dedicated and specialized service-wide readiness infrastructure that matches resources with requirements, and attains and maintains readiness to facilitate rapid activation and deployment of the USCGR when surge operations require additional personnel for the active component."[4]

---

[2] "The Integration of Active and Reserve Forces," *Coast Guard Reservist*, October 1994.

[3] Discussion conducted at Coast Guard headquarters with senior USCGR officials, August 25, 2016.

[4] See USCGR, "Workforce Organization," undated.

A key part of this system was the creation of the positions of senior reserve officer and senior enlisted reserve adviser. There are very few designated command and leadership positions in the USCGR since the integration. The new role of the USCGR is to augment the active Coast Guard and to essentially be viewed as trainees working to achieve higher levels of competencies. This perception of being trainees applies even to the O4/O5 level because these officers are training to be ready to augment the active Coast Guard for crisis response and surge requirements. Although this would be a foreign idea to the other services' RCs, it is a cultural norm for the Coast Guard.

That said, the desire to lead runs strong among the officer corps and sometimes creates situations that ultimately result in readiness challenges. The scenario runs something like this: The senior reserve officer in a given context does not want his or her reserve unit to be a burden on the active Coast Guard and desires to lead. The active Coast Guard unit commander is both sensitive to that and has a lot on his or her plate. So a "shadow command" situation is created, which seems like a win–win for both. The unfortunate result, however, is that the active Coast Guard commander then does not take full ownership and responsibility for the USCGR unit's readiness, and the senior reserve officer focuses more on leading and management to the exclusion of earning his or her required competencies, which is how readiness is measured in the Coast Guard. This is a good illustration that lasting change takes time to fully take hold and that, often, resistance can stem from good intentions.

Current Coast Guard Models of Integrated Organizational Structures and Processes

Figure 6.1 shows the overall organization of the Coast Guard. Coast Guard units fall under two areas: Pacific Area and Atlantic Area. The Pacific Area is divided into four districts, which are further divided into a total of 11 sectors. The Atlantic Area is divided into five districts, including a total of 26 sectors.[5] About 65 percent of the USCGR is attached to the sectors. In this structure, the senior reserve officer positions end up as among the most sought after in both the AC and RC. Each component has O6 positions in the sectors—an active Coast Guard commander and a senior reserve officer. Achieving an O6 senior reserve officer position is now considered representative of a very successful USCGR career for an officer.

The Coast Guard does not have a position for chief of the USCGR, nor a separate ADCON chain of command for the USCGR. Figure 6.2 expands the organization of the assistant commandant for human resources (CG-1) under the DCMS. Within CG-1 is the Reserve and Military Personnel Directorate (CG-13). Some key points from the mission statement and functions of CG-13 are as follows:

- Provide the Coast Guard a ready reserve force that embodies the competencies necessary to perform maritime homeland security.
- Provide an active-duty force to support all missions of the service.
- Serve as the commander of the Coast Guard RC and program director of the reserve training program.
- Develop a reserve program vision and strategic guidance for the USCGR in alignment with the Coast Guard's Reserve Policy Statement.
- Assist program managers to determine surge, mobilization, augmentation, and other part-time workforce requirements that make the best use of full- and part-time resources.
- Develop workforce management plans and policies; ensure overall policy consistency between the AC and RC of the Coast Guard.[6]

There are three important points to highlight from the organizational structure, mission statement, and functions of CG-13. The most striking point is that the active and reserve military personnel functions are consolidated within a single directorate. That alone highlights

---

[5] U.S. Coast Guard, "Units," last modified July 5, 2017.

[6] See DCMS, U.S. Coast Guard, U.S. Department of Homeland Security, "Reserve and Military Personnel Directorate (CG-13)," undated.

Figure 6.1
Overall Organization of the Coast Guard

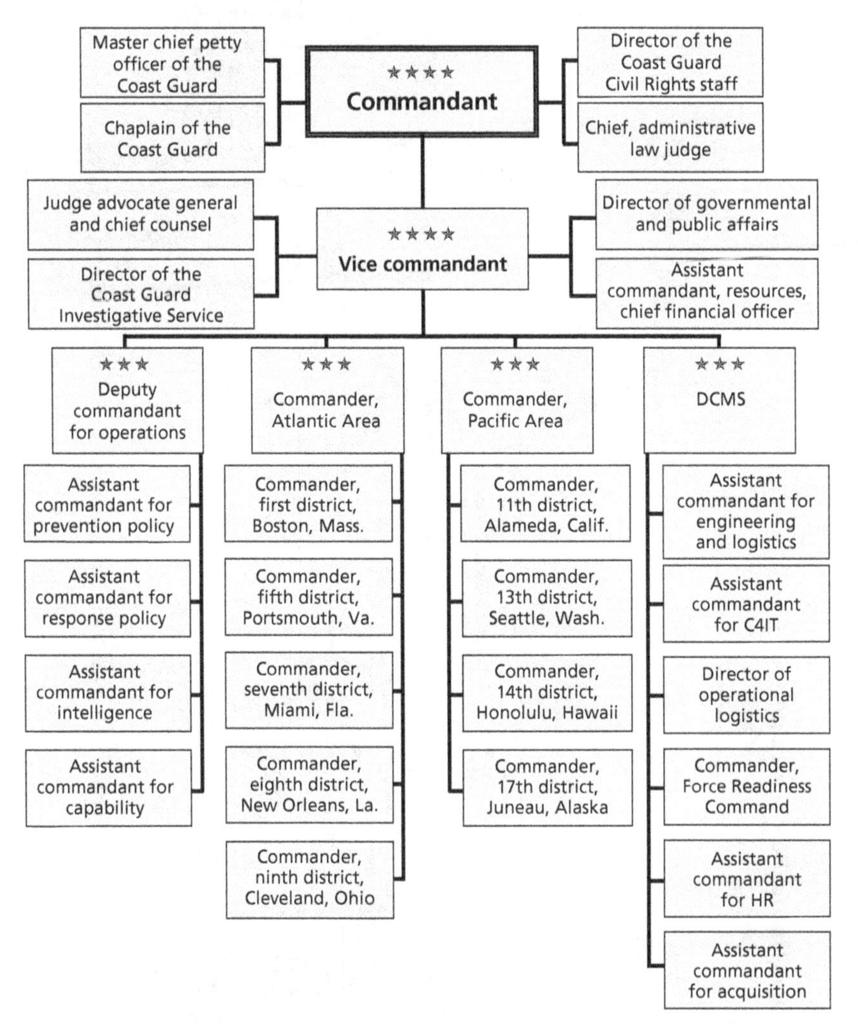

SOURCE: U.S. Coast Guard, *U.S. Coast Guard Overview*, Washington, D.C.: U.S. Coast Guard Headquarters, CG-PTT, October 2016.
NOTE: C4IT = command, control, communications, computers, and information technology. HR = human resources.
RAND *RR1869-6.1*

Figure 6.2
Coast Guard Headquarters Human Resources Directorate

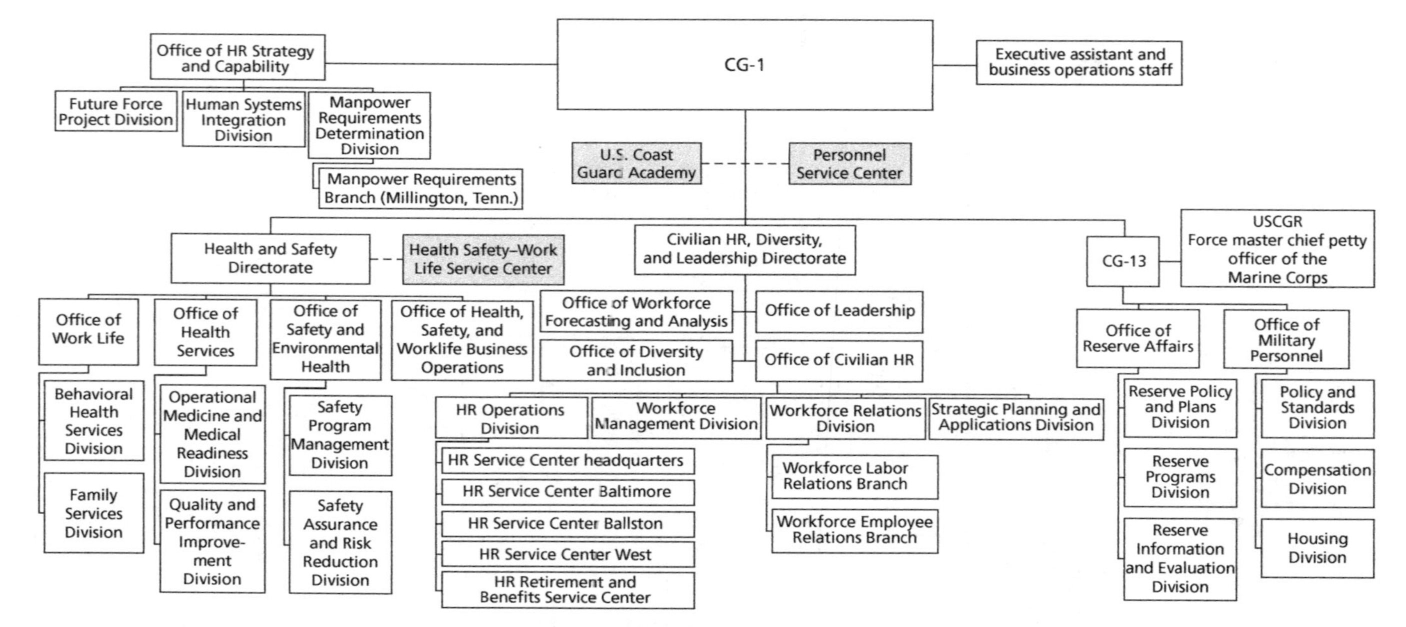

SOURCE: DCMS, organizational chart for the Office of the Assistant Commandant for Human Resources, updated October 17, 2016.

RAND *RR1869-6.2*

the degree of integration between the AC and RC. The second is that any advocacy specifically for the USCGR is for the purpose of ensuring its readiness to augment the active force as needed, both operationally and for mobilization. Third, the directorate works to "make the best use of full and part-time resources."[7] Although this last point is related to the second point above, it is an additional nuance that emphasizes both the total force nature of the Coast Guard and the USCGR's role as a provider of part-time forces to the Coast Guard rather than as an independent entity. Something that makes the integration of the AC and RC of the Coast Guard even more striking is that the commander of the Coast Guard RC is most often an active-duty admiral.

Lessons Learned

Up until 1994, the two components of the Coast Guard operated separately but with shared equipment. With the goal of generating maximum resources from the AC and RC of the Coast Guard, in 1994, the commandant of the Coast Guard tasked a working group to explore integration. He then directed integration of the active Coast Guard and the USCGR, with the very clearly defined end state described in this chapter. This clear vision and a supporting implementation plan appear to have greatly facilitated the Coast Guard's 1994 integration effort.

As a result of this effort, the USCGR is integrated into active Coast Guard day-to-day operations. This has been made possible by the integrated training of AC and RC personnel. Because RC personnel are now directly assigned to the active units in which they train, their efforts and skills are directly applied to the unit's mission.

One of the major challenges with the Coast Guard's integration efforts was the loss of leadership positions in the USCGR. Although this was clearly a perceived loss for many, Coast Guard leadership remained firm in pursuing the vision of one total Coast Guard team under active Coast Guard unity of command.

---

[7] See DCMS, undated.

Assessment of Active and Reserve Component Integration in the Coast Guard

The Coast Guard's integration efforts reinforce the importance of the best practices and success factors identified in Chapter Three. For instance, the goal of the 1994 integration effort had a very clearly defined end state, and the vision was communicated in multiple ways, among them in the October 1994 issue of *The Coast Guard Reservist*, which stated,

> From the classroom meetings on Thursday nights in the 1950s, to the "separate but equal" practice of augmentation training of the 1970s and 80s, we have come finally to integrating our reserve into the operating missions and administrative processes of the regular Coast Guard . . . the result is team Coast Guard.[8]

Our research does not indicate any attempt to build a coalition to support the integration effort—the change was directed and implemented swiftly. The Coast Guard also linked the implementation of the integration effort to this vision. In particular, the new structure of the Coast Guard was directly tied to the vision of the 1994 integration effort.

The Coast Guard has very clearly stated roles and expectations for its AC and RC. Fulfilling these roles and expectations is achieved through structural and organizational alignment and staff processes. One of the primary challenges to achieving enduring success is cultural. A service can always implement structural, organizational, and policy changes, but challenges can remain if the organizational culture resists the changes.

A 1996 study noted both the promise and the cautions for the direction on which the Coast Guard embarked in 1994:

> The very act of integration has already removed many of the barriers in the Coast Guard between the active and reserve components, and in the operating units both regulars and reservists

[8] "The Integration of Active and Reserve Forces," 1994.

> are working together for the common good with a minimum of fuss and turmoil. The Coast Guard still has some problems to solve, but they are relatively minor compared to the problems that existed before integration.[9]

The report goes on to say, "The Coast Guard has achieved structural integration and administrative integration. It remains to be seen whether it can also achieve cultural integration."[10]

One of the major challenges associated with the Coast Guard's integration effort was the loss of USCGR leadership positions. Although there was initially some attrition due to this loss of leadership positions and the Coast Guard continues to wrestle with the lack of formal command and leadership opportunities, Coast Guard culture has evolved into a total force culture. Overall, our findings indicate that the Coast Guard has adapted quite well to integration post-1994.

---

[9] John R. Brinkerhoff and Stanley A. Horowitz, *Active–Reserve Integration in the Coast Guard*, Washington, D.C.: Office of the Assistant Secretary of Defense for Reserve Affairs, October 1996, p. 10.

[10] Brinkerhoff and Horowitz, 1996, p. 10.

CHAPTER SEVEN

Active and Reserve Component Integration in the Marine Corps

The AC and RC of the Marine Corps are highly integrated and arguably represent the best tested example of total force integration among the case studies considered. The Marine Corps consists of an AC and a federal reserve—the Marine Corps does not have a National Guard. The authorized end strengths of the Marine Corps's AC and RC are 202,100 and 39,600, respectively.[1] The breakout of MARFORRES is shown in Table 7.1. Excluding the IRR, MARFORRES is 18 percent of the total force.

Evolution of Active and Reserve Component Integration in the Marine Corps

What the Marine Corps has settled on to date with regard to RC integration reflects many years' worth of different organizational concepts.

---

[1] U.S. Marine Corps, "Headquarters Marine Corps," home page, undated. The Selected Marine Corps Reserve (SMCR) is the largest subcategory of the Ready Reserve—the category of reservists most often called to active duty. SMCR members are required to attend training and are available for recall to active-duty status. The apparent discrepancy between the numbers cited for the RC in the first paragraph above—39,600—and the number cited for the U.S. Marine Corps Forces Reserve (MARFORRES) is explained as follows. The overall MARFORRES number includes both the Individual Ready Reserve (IRR) and the assigned AC, neither of which is considered part of the authorized reserve end strength. The total of the remaining elements of the MARFORRES shown in Table 7.1 constitute the selected reserve, and that number is 38,413. The remaining discrepancy stems from variations in initial active-duty training numbers and when the data were cited.

Table 7.1
Marine Corps Forces Reserve End Strength

Segment	Number of Personnel
IRR	69,207
Selected Marine Corps Reserve	30,519
AC	4,025
Initial active-duty training	3,140
IMAs	2,528
Active Reserve	2,226
Total	111,645

SOURCE: Office of MARFORRES, 2016.
NOTE: IMA = Individual Mobilization Augmentee.

In 1925, after the passage of an act to create the Marine Corps Reserve,[2] an independent reserve section was organized within the office of the commandant of the Marine Corps to perform reserve functions. Prior to that, reserve functions were distributed across the staff. This reserve section was expanded in 1937 to a reserve division, reflecting an increase in the number of functions being performed. Also during this time period, the commandant requested that the Department of the Navy provide a recommendation for reorganizing the Headquarters, Marine Corps (HQMC) staff. One of the results of this was the creation of a personnel department in 1943 that absorbed the reserve division. After World War II, as a result of analysis of lessons learned, Reserve Affairs was separated from the personnel division as part of a larger effort to rebuild the RC. This Reserve Affairs Division eventually became a separate deputy chief of staff element (reporting directly to the commandant), and this organization remained until 1988. In this time period, the Marine Corps also established the 4th Marine Division and the 4th Marine Aircraft Wing (both Marine Corps reserve

[2] Public Law 68-512, An Act to Provide for the Creation, Organization, Administration, and Maintenance of a Naval Reserve and a Marine Corps Reserve, February 28, 1925.

organizations), and eventually created a headquarters to exercise command and control over these reserve forces.

In 1988, a Manpower and Reserve Affairs Department was created, and the Reserve Affairs Division was integrated into and across that larger department but ceased to exist as a separate organization. Where previously the director of the Reserve Affairs Division advised the commandant directly on RC matters, in the new department, the director was an adviser to the new deputy commandant for manpower and reserve affairs, who then became the direct adviser to the commandant on reserve issues. This new arrangement lasted only three years because, after receiving complaints from MARFORRES personnel, the Senate Armed Services Committee directed the Secretary of Defense to improve oversight and responsiveness to RC matters at the service headquarters level, and, as a result, a Reserve Affairs Division was reestablished within the office of the deputy commandant for manpower and reserve affairs.

The evolution continued in 1996 with the establishment of the Office of MARFORRES, headed by the commander of MARFORRES (COMMARFORRES) and charged to be the commandant's principal adviser on MARFORRES matters. This situation makes the Marine Corps different from the Army, Air Force, and Navy, in that the reserve chief in each of those services is dual-hatted as the commander of that service's respective reserve forces. So although COMMARFORRES commands the reserve units, in the Marine Corps, there is no "service chief reserve office." Although the Office of MARFORRES proposed a change to create a service chief reserve office, the Marine Corps currently manages its RC through the following command structures:

- The deputy commandant for manpower and reserve affairs is the principal staff officer for reserve manpower matters and is directly responsible for the formation of plans, policies, budget, structure, and administration of the RC.
- The director of the Reserve Affairs Division is the principal adviser to the deputy commandant for manpower and reserve affairs on all matters pertaining to the RC.

- COMMARFORRES is the principal adviser to the commandant on matters pertaining to the MARFORRES.[3]

The Marine Corps consciously chose the current organization described above to better enable it to truly operate as a total force. What history shows is that, since 1925, HQMC has gone through several iterations and evolutions of responsibility for RC matters, being totally separate from the AC, being totally integrated within the AC, and the balance represented today that is different from the other services.

Current Marine Corps Models of Integrated Organizational Structures and Processes

In the Marine Corps's organizational design, MARFORRES is an integral element. The MARFORRES organization is integrated into the chain of command going up to the commandant of the Marine Corps. For instance, from a process perspective, COMMARFORRES is a full voting member on the Marine Requirements Oversight Council. This is significant because this council advises the commandant of the Marine Corps on policy matters related to concepts, force structure, and requirement validation.[4] Also, the billet for COMMARFORRES is an AC billet. There is no expectation that COMMARFORRES will always be filled by a reserve general officer. Of the past six COMMARFORRESes, including the current incumbent, LtGen Rex C. McMillian, who is RC, three have been AC and three have been RC. This is yet another statement attesting to the integration of the two components.

Figure 7.1 illustrates how the AC and RC organizational structures compare and indicates that the AC and RC structures are almost

[3] The information in this section was drawn extensively from K. J. Conant, U.S. Marine Corps Reserve, "History and Overview of Reserve Affairs Division," August 26, 2014.

[4] Commandant of the Marine Corps, "Marine Requirements Oversight Council," Memorandum 1-02, January 17, 2002.

Figure 7.1
Marine Corps Active and Reserve Component Structure Comparison

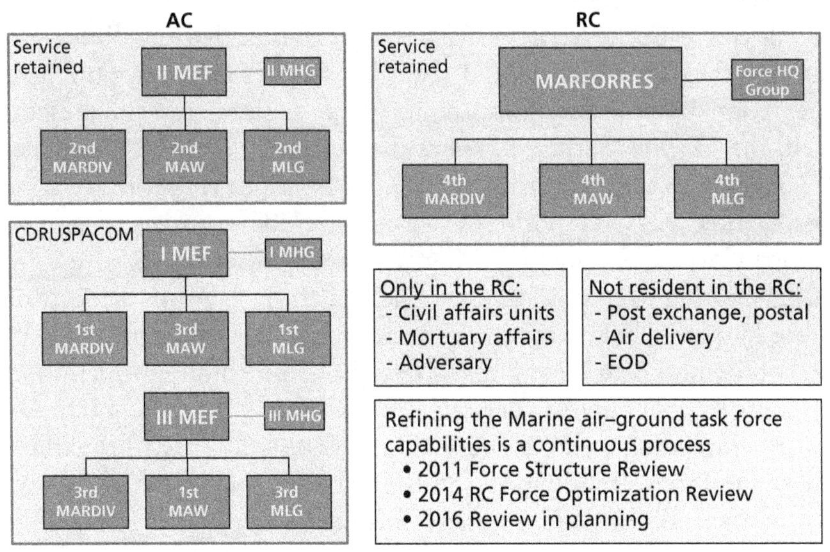

SOURCE: Andrew Ryan, deputy, strategic communications officer, Office of Marine Forces Reserve, "Total Force Integration Brief," briefing slides from briefing to MG (ret.) G. A. Schumacher, U.S. Army, last updated June 28, 2016.
NOTE: MEF = Marine expeditionary force. MHG = Marine headquarters group. MARDIV = Marine division. MAW = Marine aircraft wing. MLG = Marine logistics group. EOD = explosive ordnance disposal.
RAND *RR1869-7.1*

the same. The only difference, as we explain below, is that the RC organization has a large number of AC marines embedded.

Regarding integration on the AC side, HQMC has a small number of active reservists embedded in the headquarters staff sections to provide RC perspective as total force policies and actions are developed and executed. It is important to note that these embedded active reservists are not expected to be proponents for MARFORRES. Rather, they are expected to be advocates to ensure that RC realities, timelines, and nuances are fully considered.[5] Some IMAs are also assigned to augment HQMC when required. The same is true for the

5 Discussion with a senior Marine Corps official in the Reserve Affairs Division, August 23, 2016.

other AC organizations in the Marine Corps, such as U.S. Marine Corps Forces Cyberspace, U.S. Marine Corps Forces Pacific, and U.S. Marine Corps Forces Central Command.

It is within the MARFORRES organization that the differences between the Marine Corps and the other services in multicomponent integration become more apparent. When a given mission presents itself, the Marine Corps will select the best units available—whether AC or RC—to execute the mission. What is unique is how the Marine Corps integrates AC positions within the RC organizations. The Marine Corps sees value in having some form of formal AC oversight and mentoring to improve the readiness of the RC. What the Marine Corps has done, both structurally and through its inspector-instructor (I&I) program, is to embed that oversight and mentoring function in the units by assigning AC personnel to RC units. There are 4,025 AC marines providing training and readiness oversight for RC formations and organizations totaling 35,885 RC marines, for a ratio of one AC marine to every nine RC marines. This ratio seems to suggest significant training and readiness oversight impact.

Another point worth noting, which might be more profound from a total force perspective, is the integration of AC marines into the command and control structure of the MARFORRES. The 4,025 AC marines mentioned above are considered in the overall MARFORRES end strength because they are actually a part of the MARFORRES units and structure. Not only are the AC I&Is embedded into the actual unit and organization structure, but they are also part of the same chain of command. When he was commanding general of the MARFORRES, MajGen Thomas L. Wilkerson stated that "the purpose of the integration was to foster a single unit identity. Therefore, a reserve regiment or unit is referred to as a [Marine Corps] unit, not a [Marine Corps] reserve unit."[6] Keeping in mind the layering of AC and RC at subsequent command levels, the majority of the I&Is will report to the major subordinate command (division, Force Headquar-

[6] James S. Santelli, *A Brief History of the 4th Marines*, Washington, D.C.: Historical Division, Headquarters, U.S. Marine Corps, Marine Corps Historical Reference Pamphlet, 1970, p. 115.

ters Group, MAW, and Marine logistics group) chief of staff, who is AC. The battalion-level I&I will report to the major subordinate command commanding general, who is RC.

Finally, the Marine Corps treats AC assignments into RC formations as an operational tour.[7] This is significant because these types of assignments are considered to be career enhancing. This type of incentive can go a long way in fostering cross-component integration.

Lessons Learned

The Marine Corps model seems highly integrated. The Marine Corps is a multicomponent organization with a single chain of command. It has a single rating scheme across components; a single, integrated pay and personnel system (which has been in place for more than 25 years);[8] a single purpose; and integrated ADCON. AC reports to RC and RC reports to AC, all within the same organization, operating as an integrated total force. The Marine Corps model appears free from the conflict and competition between the components in some of the other services.

As is discussed elsewhere in this report, unity of command is a defining characteristic of a true total force consisting of forces from both AC and RC organizations. Although the separate components bring their different strengths to the overall organization, unless a single entity makes the final decisions about equipping, manning, and utilization, the result is competing components vying for their own interests. Consider the mission statement of the MARFORRES:

> Commander, Marine Forces Reserve (COMMARFORRES) commands and controls assigned forces for the purpose of augmenting, reinforcing, and sustaining the AC with trained units and individual marines as a sustainable and ready operational reserve in order to augment and reinforce active forces

[7] Discussion with a senior Marine Corps official in the Reserve Affairs Division, August 23, 2016.

[8] Promotion boards remain separate.

> for employment across the full spectrum of crisis and global engagement. On matters pertaining to Marine Forces Reserve, COMMARFORRES serves as the principal advisor to Commandant of the Marine Corps.[9]

This emphasizes that the MARFORRES' role is to support the AC and to maintain a ready, relevant, and responsive force that provides the AC with "a shock absorber"[10] when requirements dictate. The MARFORRES mission statement, along with the well-defined chain-of-command lines, make it clear that the MARFORRES is not only a complementary force to the AC but also ultimately subordinate to the AC leadership (through unity of command) as part of one, integrated total force.

Assessment of Active and Reserve Component Integration in the Marine Corps

The Marine Corps's integration efforts reinforce the importance of the best practices and success factors identified in Chapter Three. The Marine Corps followed a more evolutionary path to integration than the other services. Although there was evolution in the reserve force structure, the changes occurred primarily at the Marine Corps staff level. At that level, incremental changes occurred by assessing progress and adjusting accordingly several times between 1925 and 1996. Sometimes, the champion of change was the commandant. At one point, it was the Senate Armed Services Committee. But since 1996, the structure and integration at HQMC have remained constant, with responsibility for the Marine Corps RC balanced between the AC and RC. There is clearly a vision of how the total Marine Corps operates, and how the Marine Corps RC and MARFORRES operate within that vision. According to our discussions with Marine Corps leader-

---

[9] Richard P. Mills, lieutenant general, U.S. Marine Corps, commander, "Vision of the Commander, Marine Forces Reserve," undated.

[10] Mills, undated.

ship, it would be inconceivable for the RC to mount a challenge to the AC based on some perception of inequitable treatment.

The Marine Corps seems to have achieved its vision of total force integration. This might explain why we found no evidence of coalition building and no plan for further change within the Marine Corps: It would not be necessary at this point. The vision of one Marine Corps, with the Marine Corps RC and the MARFORRES existing to support and augment the active Marine Corps as required, is a consistent theme. The lack of a National Guard component means that there have been no Title 32 legal issues with which to contend. The practices regarding culture and momentum are apparent: There is a very distinct total Marine Corps culture that is maintained, and reserve marines are considered simply marines. Although there have been changes in the structure of the HQMC staff, they seem to have been refinements that resulted in a balance in which reserve-specific issues are represented within a unified command structure.

CHAPTER EIGHT

Active and Reserve Component Integration in the Navy

The U.S. Navy consists of an AC and a federal RC. The Navy does not have a National Guard component, although some states have naval militias.[1] As indicated in Figure 8.1, the authorized end strengths of the Navy's active personnel and Selected Reserve (SELRES) personnel are approximately 328,500 and 58,000, respectively.[2]

Evolution of Active and Reserve Component Integration in the Navy

Beginning in 2003, the Navy embarked on a path of integrating its active and reserve forces into a holistic Total Navy force that habitually and routinely calls on its RC to support ongoing operations. This integration could be considered radical in that the Navy went about it with a sense of urgency after an internal study concluded that the Navy Reserve was not structured properly, and there was a perception that

[1] Naval militias are somewhat similar to the National Guard but have differing degrees of federal recognition and obligations. Some militia members are also members of federal reserves and have served in Iraq and Afghanistan. Only a very few states have a naval militia, and the number of personnel is fairly small. See Albert A. Nofi, *The Naval Militia: A Neglected Asset?* Alexandria, Va.: CNA Corporation, CIM D0015586.A1/Final, July 2007. Nofi describes three models of naval militias (pp. 34–36). Because of the small numbers and the varying levels of federal recognition, our study did not consider them.

[2] These numbers also include FTS personnel.

Figure 8.1
Navy Total Force End Strength

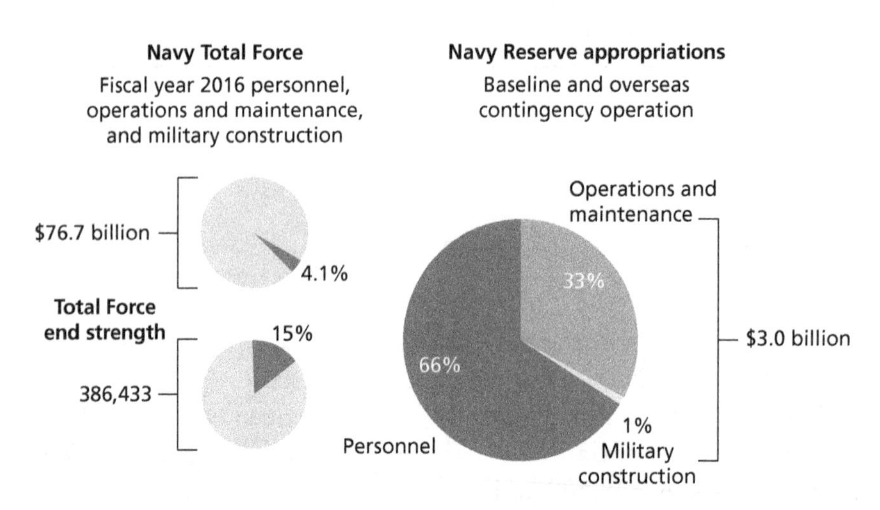

SOURCE: Commander, Navy Reserve Forces (COMNAVRESFOR), 2016.
RAND *RR1869-8.1*

the Navy had not fully embraced its reserve.[3] VADM John G. Cotton, the Chief of Navy Reserve, whom then–Chief of Naval Operations (CNO) ADM Vernon E. Clark charged to execute this vision, was expected to produce major changes within a year. The Navy believed that revolutionary change was required. The end result is a much more integrated total Navy, with a common set of standards not present prior to 2003.

According to our discussions, the Navy believes that it has achieved the correct balance between philosophical integration and actual, physical integration at the headquarters staff and unit levels—a balance that served the Navy well through Operations Enduring Freedom and Iraqi Freedom. The philosophical integration lays the framework for the goals and end state of the integration, which, in turn, drives organizational structure and the degree to which physical inte-

---

[3] Steve Keith, "From the Editor," *Naval Reserve Association News*, Vol. 51, No. 5, May 2004, p. ii, reporting on an interview with William A. Navas Jr., Assistant Secretary of the Navy for Manpower and Reserve Affairs, who reported his conversation with VADM Thomas J. Ryan Jr. in early 2004.

gration between the components is required to achieve the institution's goals. The Navy decided on a clear vision of what its total force should be and how it should operate, which, in turn, drove the roles of the two components with respect to one another and the resultant organizational structures to best achieve that vision.

As mentioned above, the Navy set about achieving a new level of AC/RC integration in 2003. Navas, then Assistant Secretary of the Navy for Manpower and Reserve Affairs, described the challenge of the transformation as "requiring a fundamental change in the way we manage, employ and even think about Reserve personnel and their units."[4] He described the two biggest changes as follows: "Navy Reservists are focused on the priorities of the operational Fleet, and the AC of the Navy is assuming responsibility—*and ownership*—of their training, recruiting, equipping, and readiness as part of seamless integration."[5] That seamless integration resulted in the Navy developing a human capital strategy and integrated pay and personnel systems to enable greater agility and flexibility, managing the total force holistically:

> We want the right people with the appropriate skills and experiences, assigned to validated work requirements, when and where they are needed. To achieve this, *our human resource systems must manage our personnel as a total, integrated force—active duty, Reserve, civilians, and contractors*—and our people must be ready to meet the challenge.[6]

In 2003, further integration efforts were made when the COMNAVRESFOR staff was merged with the staffs of the commanders of Naval Air Reserve Forces and Naval Surface Reserve Forces to operate as a combined staff serving three separate operational commanders.

---

[4] William A. Navas Jr., "Integration of the Active and Reserve Navy: A Case for Transformational Change," *Naval Reserve Association News*, Vol. 51, No. 5, May 2004, pp. 11–19, p. 11.

[5] Navas, 2004, p. 11. Emphasis ours.

[6] Navas, 2004, p. 12. Emphasis ours.

The ground work for determining the best balance and mix of numbers and capabilities in the proper component was laid by an extensive Naval Reserve Redesign study headed by then–Vice CNO ADM William Joseph Fallon and then–Deputy Assistant Secretary of the Navy for Reserve Affairs Col (U.S. Army, ret.) Harvey C. Barnum Jr. The objective for that exercise was to determine how the Navy Reserve could best complement the AC Navy for operational missions. That is a far different construct from just funding increased strategic capacity in the Navy Reserve. It is one that allows for a smaller AC Navy by providing the means for both augmentation and surge capability for routine operations and emerging crises not requiring lengthy mobilizations of the RC. VADM Gerald L. Hoewing (Chief of Naval Personnel) and Admiral Cotton (Chief of Naval Reserve) summarized the change in their joint testimony before the House Armed Services Committee in 2005:

> We are completing transformation of the Navy Reserve from a strategic, standby force to an integrated and engaged operational force. As a result of ongoing Navy active and reserve integration, Navy Reserve support will increasingly take the form of individual sailors augmenting AC forces, over the traditional approach of mobilizing as entire RC units. Continuing to refine our ability to meet the "demand signal" selectively, vice en masse, will greatly ease the burden on the RC and will provide many more capabilities and opportunities for Reservists to serve.[7]

[7] Gerald L. Hoewing, vice admiral, Deputy Chief of Naval Operations for Manpower, Personnel, Training, and Education, U.S. Navy; and John G. Cotton, vice admiral, Chief of Navy Reserve, U.S. Navy, joint statement before the U.S. House of Representatives Committee on Armed Services Subcommittee on Military Personnel on recruiting and retention, July 19, 2005, p. 7.

Current Navy Models of Integrated Organizational Structures and Processes

Figure 8.2 illustrates different models of scalable support options for the Navy Reserve and examples of associated unit types.

The first of these models, the augmentation model, consists of units that are specifically designed to provide an augmentation capability to AC Navy units. They are not stand-alone units that would mobilize or deploy to perform a specified mission as a unit. This available pool allows for a smaller AC Navy force, with a ready source to draw from for manning shortfalls and additional surge capacity for operational missions or emergent crises. This model accounts for 71 percent of the Navy's SELRES.

The second model is the special capability model. These represent important niche capabilities (such as submarine rescue and the EOD Technology Division) that the Navy has resident only in the Navy Reserve.

The third model, blended units, represents a model that other services are using and with which they are experimenting. These units have core AC personnel, but using the unit's full capacity and capability requires that RC members be brought into an active capacity through either mobilization or a shorter-term duty status, such as active duty for training or active duty for special work.

The final model, the component unit model, represents the more traditional strategic role of the RC. These are units that mirror AC units and therefore provide the total force with a larger mobilization capacity of like units at a lower carrying cost. This model, along with the special capability and blended unit models, represent the remaining 29 percent of the Navy's SELRES.

In order to accomplish the two major cultural and organizational shifts mentioned above—integrating the Navy Reserve into the support of the ongoing operations and making AC Navy commanders responsible for the training, recruiting, equipping, and readiness of

Figure 8.2
Integrated Navy Reserve Utilization

Augmentation models	Special capability models	Blended unit models	Component unit models
Added operational capacity	Niche skills	AC/RC units	RC units
• Command staff augment and exercise support • Maritime operations center watch standers • Naval Security Force • Naval hospital support • Religious ministry • Training teams • Surface force and submarine force readiness units • Office of Naval Research, Naval Air Systems Command, Naval Sea Systems Command, Space and Naval Warfare Systems Command	• Submarine rescue (Undersea Rescue Command) • Navy emergency preparedness liaison officers • Naval Cooperation and Guidance for Shipping • EOD Technology Division • Shipyard maintenance support • High-value unit escort	• Squadron augmentation units: helicopter mine countermeasure squadron, helicopter maritime strike squadron, helicopter combat support squadron, fleet replacement squadron, training squadron, carrier airborne early warning squadron • Information Dominance Corps • Tactical support units (formerly Helicopter Combat Support Squadrons 84 and 85) • Cyber Mission Force • Unmanned systems (Triton)	• Naval mobile construction battalion • Coastal riverine squadron • Expeditionary logistics • Aviation squadrons (fleet logistics support squadron, helicopter light antisubmarine squadron, fixed-wing patrol squadron, tactical electronic warfare squadron, strike fighter squadron, fighter squadron composite) • Sea, air, and land teams

Scalable mobilizations: Individual augmentees or units in support of Global Force Management requirements

Capabilities flexibly structured to meet Navy requirements

SOURCE: COMNAVRESFOR, 2016.
NOTE: HSC = helicopter combat squadron.
RAND *RR1869-8.2*

the Navy Reserve—some policies were changed and mind-sets shifted. Among them were the following:

- On written correspondence, in signature blocks, the *R* was removed from Navy Reserve.
- Navy Reserve Centers were renamed Navy Operational Support Centers.
- AC and RC personnel and funding cuts are increasingly considered holistically, with the Navy recognizing that cuts to the reserve will affect overall operational capability and flexibility.
- Navy commanders became responsible for the training and readiness of the augmenting reserve units, and *the AC commander became responsible for writing the fitness report for the augmenting RC commander.* These changes in particular illustrate the extent of the Navy's total force integration.[8]

Lessons Learned

From the start of its integration efforts, the Navy had a clear vision of what end state it wanted to achieve—a conscious move away from a purely strategic reserve to a holistic, total force—and it designed its RC to be primarily a scalable and flexible augmentation force for the AC. This clarity provided benchmarks and metrics for success. Because each service is very different in terms of structure, size, types of units, how they are employed, and so on, this is not to suggest that the Navy model should, or even could, be reflexively applied to the other services. However, the key principles and processes the Navy employed during its integration do apply to any of the other services contemplating or

[8] These policy changes and mind-set shifts were described in discussions with senior Navy Reserve personnel from the Navy Reserve Forces Command and the Office of the Chief of Navy Reserve, on September 13 and 15, 2016, respectively. Emphasis is ours.

experimenting with different forms of AC/RC integration. Those principles are as follows:

- clearly defined end state
- zero-based review to determine optimal force mix and organizational structures
- wherever the service chooses to implement integration between its components, unity of command and a single rating scheme
- deliberate but urgent implementation.

Some senior Navy leaders believe that optimal AC/RC integration is a matter of finding the right balance. In other words, integration should not be pushed as far as possible simply for its own sake; instead, integration should occur to the degree that is necessary to achieve the organization's objectives and not beyond. For example, when Admiral Cotton executed full-scale integration in 2003–2004, some RC FTS personnel were embedded into the Office of the CNO staff, with the intent that they focus on and represent RC issues and perspectives to the functions of the staff section and to the AC principals. But what ultimately happened is that they became subsumed as more action officers on the AC staff. This created the potential for the advocacy for the RC position, challenge, or issue to become diluted or even lost because of the military rank structure.

This example suggests that a fully knowledgeable and rank-equivalent advocate is necessary to ensure that RC challenges and issues are fully considered and advocated for. Further, some senior Navy leaders believe in the efficacy of how the Navy has split operational control down the AC chain to its supporting RC units and ADCON down the Navy Reserve Force (NAVRESFOR) chain of command. They see this not as duplicative but as a means of most efficiently preparing the RC sailors and units to be ready assets that the AC can employ operationally. Although it would be possible that, over a long period of time, the AC commanders could gain sufficient expertise to render the separate RC ADCON chain truly duplicative, this might not occur any time soon. Despite increasing familiarity with the RC and education on RC issues and challenges, aspects remain that they do not fully grasp, just as there are areas in which the AC has knowledge and experience that the RC does not possess. Therefore, it is critical to ensure that the right people with the right knowledge, background, and experiences are in the appropriate positions to fully execute and represent their areas of expertise such that the total force operates in an optimal manner.

Assessment of Active and Reserve Component Integration in the Navy

The Navy's integration efforts reinforce the importance of the best practices and success factors identified in Chapter Three. Like those of the Coast Guard, the Navy's integration efforts in 2003 represented a major shift from the status quo that took place at the direction of leadership. The need and the vision were very clear: to manage all Navy personnel as an integrated, holistic force in order to accomplish the priorities of the operational fleet. Like in the case of several of the services, there was no coalition building, only an order to execute based on the results of an exhaustive "bottom-up review" to determine how the Navy Reserve could best complement the active Navy for operational missions.

Some of the most-important aspects of the implementation of the Navy's integration efforts were instituting unity of command and making the active Navy responsible for the recruiting, training, equipping, and readiness of the RC. Therefore, the resulting structural and policy changes were both congruent with and necessary for the achievement of the vision. Other changes that were important to cultural change and maintaining the momentum of the integration efforts included the development of an integrated pay and personnel system, removal of "Reserve" from correspondence, and AC commanders writing RC fitness reports. These have all kept the force holistic in outlook. Findings from our discussions with Navy leadership indicate that the integration went very well and that the changes postintegration have been widely accepted and deemed to have been both needed and effective. Those discussions also indicate that much of the success was due to a ready supply of overseas contingency operations funding for the operational employment of the Navy Reserve. With Operations Enduring Freedom and Iraqi Freedom concluded, we were told, the Navy will again need to examine the best role for the Navy Reserve, the AC/RC force mix, and the level of integration to determine whether the result of the 2003 integration effort remains the best model and, if not, to establish clear objectives for possible alternatives.

CHAPTER NINE

Findings and Recommendations for Future Integration Efforts

The review of the management and organizational literature presented in Chapter Three revealed several factors—including approaches and practices for undertaking and implementing integrations, potential barriers, ways to overcome barriers, and pitfalls to be avoided—that can affect the likelihood of integrating organizations successfully, of implementing organizational change successfully, and of making organizations generally more successful. From that list of factors, we distilled a set of best practices that can be helpful when undertaking and assessing efforts to integrate organizations. In Chapters Four through Eight, we described the case studies of AC/RC integration efforts in the U.S. Air Force, Army, Coast Guard, Marine Corps, and Navy. The first section of this chapter presents the findings from those case studies in the framework of the best practices. It also includes several recommendations relating to some of the practices that might help to improve the chances of success in future integration efforts.

The case studies revealed some additional best practices that are more specific to integrating AC and RC organizations. We describe those practices and the related findings from the case studies in the second section of the chapter, which also includes recommendations associated with those practices.

Best Practices as Reflected in the Case Studies and Recommendations for Future Integration Efforts

The best practices from the literature described in Chapter Three were present to varying degrees in the cases of AC/RC integration that we examined, and we describe those findings in this section. Two things should be kept in mind. First, we did not always find enough information when examining the case studies to determine the extent to which a practice was followed. Second and most importantly, we present this information in case it can help future AC/RC integration efforts to improve the chances of success. In presenting it, we do not intend to judge how well any service has carried out its integration efforts or whether one service has integrated better than another.

As we noted in Chapter One, there have been many efforts to achieve various degrees of integration among active, reserve, and guard forces in past decades, and these are almost certain to continue. Some efforts can be driven from inside the military or DoD—to fix particular problems that leaders perceive or to reduce manpower or spending—while others can come from outside sources, through legislation or commissions, such as the NCSAF and the NCFA. Whatever the source or reason, our research suggests that there are ways to improve the odds of these efforts succeeding.

All of the best practices described in this report are found to be important by the literature, our case studies, or both. But a subset of the practices stood out to the team as especially relevant to AC/RC staff integrations. We believe that, by focusing on this subset of practices, leaders can increase the chances of future integrations succeeding. We have included recommendations in this section associated with those practices.

Establish the Need and the Vision for Change

One of the most-important lessons learned from the case studies is the importance of clearly establishing the need and vision for change. This was perhaps most clearly demonstrated in the Coast Guard and Navy case studies. When the Coast Guard set out to integrate its AC and RC in 1994, it set a clear goal to generate maximum resources from the AC and RC of the Coast Guard, and it articulated its vision for an integrated Coast Guard. From the outset, the Navy had a very clear vision of what end state it wanted to achieve—a conscious move away from a purely strategic reserve to a holistic, total force—and it designed its RC to be primarily a scalable and flexible augmentation force for the AC. This clarity provided benchmarks and metrics for success.

The Air Force and Army cases also reinforce the need for a clear vision for change. The Air Force is moving forward developing plans and testing various forms of organizational structures, complete with goals and measures, but has not yet developed a clear vision for what its total force will look like and how to implement it. Like the Air Force, the Army is pursuing new approaches to integration but lacks a vision with a clearly defined end state. Like with the Air Force, there is general acceptance in the Army that different force-mix ratios are likely required to balance current readiness with future modernization, but it is unclear as to what form that should take and what degree of integration is both desirable and achievable.

Recommendation: Articulate the Need for Change and Adopt a Clear Vision for the Integration

Changing organizations, especially large ones, is difficult. People tend to be more willing to support change if they understand why it is needed and what it will look like. A clear articulation of both the need and the vision for an integration effort can serve several purposes. It can give members of the organizations a greater sense of certainty about the future. Uncertainty can undermine support, strengthen opposition, and reduce momentum. A clear vision also provides a way to assess progress. It is hard to overstate the importance of establishing a clear vision. Achieving success is much easier if success is well defined.

It is also important that the vision for the integration be realistic and achievable. Adopting a vision, for example, of a service becoming seamlessly integrated and operating with full unity of command is not realistic if current legal constraints do not allow it. If the vision is not achievable, people might not bother trying, and opposition can grow.

Create a Coalition to Support the Change

In our case studies, the discussions and analysis of the secondary literature provided no evidence that any of the services created a coalition to support cross-component integration efforts. This is not altogether surprising because senior leaders directed some of the integrations (particularly in the case of the Coast Guard and the Navy). In other cases, such as the Air Force, Army, and Marine Corps, integration efforts evolved over time.

Communicate the Vision

Our case-study findings reinforce the importance of this best practice. Some services have communicated their vision for change effectively through multiple avenues. For instance, the Coast Guard communicated its vision of its 1994 integration effort in multiple ways, among them in the October 1994 issue of *The Coast Guard Reservist*. Other services, including the Air Force, have communicated their visions through service policies and regulations. Our case studies indicate that the Air Force, Coast Guard, Marine Corps, and Navy have sustained their efforts to communicate their vision for integration over time, whereas the Army has not sustained its efforts to develop, articulate, and communicate its vision over time.

Recommendation: Communicate the Vision for the Integration Regularly

Integration, like other forms of organizational change, is a process rather than an event. The vision for an integration needs to be communicated early in the integration, but the communication needs to continue throughout the process. People need to be reminded of the need and of what they are working toward. And because military members rotate into new positions regularly, it is particularly important to periodically reinforce the need and vision for change.

Develop an Implementation Strategy, Including Goals and Measures

We could not identify whether any of the services developed implementation strategies associated with their integration efforts and, if so, what those strategies included. However, the importance of this practice was emphasized in the literature.

Recommendation: Develop a Strategy for Implementing the Integration That Includes Clear Goals and Measures of Success

The literature emphasizes the importance of a well-designed implementation strategy that states goals, assigns responsibility, identifies risks and mitigation strategies, and describes measures of progress and success for those goals. This is a critically important practice for DoD to consider when embarking on future integration efforts. Without developing an implementation strategy that has clear goals and concrete ways to measure progress toward those goals, it will be impossible to assess whether integration efforts are having the desired outcomes. Such an implementation strategy would also allow DoD to identify problems with implementation quickly and adjust course accordingly. This can save both time and money during implementation and can increase the odds that the changes can be sustained.

Link the Vision and the Structure

Our case studies indicate that this best practice is one of the critical potential failure points in the successful implementation of integration efforts and reinforce that articulating a vision for change is not enough. Unless that vision is linked to the organizational structure and institutionalized, integration efforts are likely to flounder. The Air Force, Coast Guard, Marine Corps, and Navy provide examples in which the vision for integration was articulated and then reinforced through structural and organizational alignment and staff processes. This has occurred in different ways, including aligning different organizational structures for associated units with the vision for integration, changing leadership positions to align with the vision, and changing service processes to mesh with the vision. The Army has not linked its vision for integration to its structure, and, as a result, integration efforts have evolved through fits and starts, and the vision has changed.

Recommendation: Ensure That the Planned Organizational Structure Is Consistent with the Vision for the Integration

Both the literature and our case studies reinforce the importance of this practice. Making the organizational structure—including not only the management reporting structure but also the functions the organization performs, the representation of the components in the organization, and even the organization's processes—consistent with the vision for the integration will help to ensure that the integration is sustainable. This practice also reinforces the need to ensure that the vision is clear and achievable.

Embed the Changes in the New Culture

Our case studies also reflect the importance of embedding changes in the new culture. The Coast Guard, the Marine Corps, and the Navy have all created new, total force cultures as a result of their integration efforts. All three of these services have institutionalized their integration efforts in the new culture, as well as in service policies and procedures. For instance, the Marine Corps and the Navy developed a single integrated pay and personnel system for both AC and RC personnel. The RC's ability to immediately supplement day-to-day operations in the Coast Guard, Marine Corps, and Navy is also a testament to the degree to which integration has been embedded in the total force cultures that these services created.

Manage the Integration of Cultures

The Coast Guard and Marine Corps cases stand out as illustrations of the importance of managing the integration of cultures. For instance, the Coast Guard has developed one of the most-integrated total force organizational cultures among the services. Prior to its 1994 integration effort, there was competition and some animosity among the Coast Guard components. Although some attrition that occurred during and immediately after the integration effort might have been personnel who disagreed with the integration effort, over time, the organizational culture in the Coast Guard has evolved into one that emphasizes a total force.

The Marine Corps has also undergone a similar organizational transformation. When he was commanding general of MARFORRES, General Wilkerson stated that "the purpose of the integration was to foster a single unit identity. . . . Thus, when a reserve regiment or unit is discussed, it is a Marine [Corps] unit, not a Marine [Corps] reserve unit."[1] Although the Marine Corps has always had a very strong sense of identity and esprit de corps, its integration efforts created a new culture in which the identity of being a marine supersedes the identity of being an active marine or a reservist. Accordingly, the Marine Corps does not have some of the same squabbles between its components over equities that some of the other services have had.

Recommendation: Work to Develop a Total Force Culture in the Integrated Organization

Similarities between the active and reserve military cultures can make it easy to underestimate the cultural challenges in AC/RC integrations. The real cultural challenge seems to be developing a total force culture in the integrated organization that considers the welfare of the total force first, rather than the welfare of the individual components. This challenge can be reflected in subtle ways (such as in attitudes that RC members are of lesser value), in more-obvious ways (such as in active service members not being rewarded for service in reserve organizations), and even in open and public fights between components over funding and perceived "equities." The Coast Guard and Marine Corps examples indicate that, if service leaders focus on and prioritize culture change, total force cultures can be developed over time.

Maintain Momentum

Our case studies indicate that the Air Force, Coast Guard, Marine Corps, and Navy have all been able to maintain the momentum of their integration efforts by institutionalizing the change and embedding it in the services' cultures and organizational structures. The Air Force and the Marine Corps have experienced long, evolutionary inte-

---

[1] 4th Marine Division Historical Detachment, *History of the 4th Marine Division: 1943–2000*, 2nd ed., 2000, p. 115.

gration efforts, and both have been able to sustain those efforts over time. The Coast Guard and the Navy, on the other hand, made discrete revolutionary changes to integrate their components, but they too have been able to sustain those changes over time. The Army has not maintained the momentum of its integration efforts. Instead, they have evolved periodically, and the vision for those efforts has changed.

Remember the Importance of People

Although this best practice is emphasized in the literature, we found few examples of its use in the case studies. Although we cannot determine whether some integration efforts failed because this best practice was not prioritized, it is clear from the literature that there are a host of potential ways that the services could mitigate opposition to integration and in fact incentivize personnel to work across components. These include potentially rewarding cross-component experience when selecting for positions and promotions, as well as treating personnel fairly and ensuring a sustainable career path across the components. Our case studies indicate that the Marine Corps treats AC assignments into RC formations as an operational tour.[2] This is significant because these types of assignments are considered to be career enhancing. This type of incentive can go a long way in fostering cross-component integration. Our analysis indicates that the Coast Guard did run up against resistance to its 1994 integration effort because it decreased the number of RC leadership positions.

Assess Progress and Adjust Accordingly

The importance of this practice is reinforced in the findings from our case studies. In particular, the Marine Corps case highlights the importance of assessing progress and adjusting over time. At the HQMC level, incremental changes occurred several times since 1925 by assessing progress and making adjustments. These incremental changes ultimately led to the evolution of a total force culture in the Marine Corps

---

[2] Discussion with a senior U.S. Marine Corps official in the Reserve Affairs Division, August 23, 2016.

and to a level of integration between the AC and RC that is arguably one of the highest among the services.

Additional Best Practices Identified in the Case Studies and Recommendations for Future Integration Efforts

The practices discussed above that were revealed during the literature survey are somewhat general in nature, applying to varying degrees to many different types of organizations. The case studies the research team undertook identified three additional best practices specific to AC/RC integration efforts:

- Establish unity of command.
- Address Title 10 and Title 32 barriers.
- Colocate AC and RC personnel in integrated organizations.

We discuss each of these practices in this section and include recommendations for future integration efforts.

Establish Unity of Command

One of the defining characteristics of military organizations is unity of command, which JP 1-02 defines as "[t]he operation of all forces under a single responsible commander who has the requisite authority to direct and employ those forces in pursuit of a common purpose."[3]

The case studies highlight the importance of establishing unity of command in organizations that integrate AC and RC even though it can be a significant challenge because of statutory and organizational structures. Each component can have its own separate chain of command that executes ADCON over its personnel. For example, Title 10 commanders cannot direct Title 32 airmen associated with them, and, even when the Title 32 airmen perform operational missions in a Title 10 status, the Title 10 commander does not execute UCMJ

[3] JP 1-02, 2010 (2016), p. 252.

authority over them because they return to Title 32 status and state disciplinary control at the conclusion of the operational mission.

To help mitigate this situation, the Air Force established the concept of OPDIR. According to Air Force Guidance Memorandum to AFI 90-1001, 2014:

> Operational Direction is defined as "the authority to designate objectives, assign tasks, and provide the direction necessary to accomplish the mission or operation and ensure unity of effort." Authority for operational direction of one component member over members of another component is obtained by agreements between component unit commanders (most often between Title 10 and Title 32 commanders) whereby these component commanders, in an associate organizational structure, issue orders to their subordinates to follow the operational direction of specified/designated senior members of the other component for the purpose of accomplishing their associated mission.[4]

Although OPDIR is not full unity of command, it might be as close as the Air Force can get given the constraints of existing laws.

Cross-component unity of command is also a hallmark of the Marine Corps case. HQMC has a small number of active reservists embedded in the headquarters staff sections to provide RC perspective as total force policies and actions are developed and executed. There are also IMAs assigned to augment HQMC when required. The same is true for the other AC organizations in the Marine Corps, such as U.S. Marine Corps Forces Cyberspace, U.S. Marine Corps Forces Pacific, and U.S. Marine Corps Forces Central Command. The MARFORRES organization is integrated into the chain of command going up to the commandant of the Marine Corps. When a given mission presents itself, the Marine Corps will select the best units available to execute the mission—whether the units are AC or RC. What is unique to the Marine Corps is how it integrates AC positions within the RC organizations. The Marine Corps sees value in having some form of formal AC oversight and mentoring to improve the readiness of the RC. What

---

[4] Moeller, 2014, p. 61.

the Marine Corps has done, both structurally and through its I&I program, is to embed that oversight and mentoring function within the units by assigning AC personnel to RC units. The Marine Corps has 4,025 AC marines providing training and readiness oversight for RC formations and organizations. Not only are the AC I&Is embedded into the actual unit and organization structure; they are also part of the same chain of command.

The Navy also set out to ensure unity of command across its components. There are different forms of associated units in the Navy with various forms of integrated command structures. In addition, the Navy has also made AC commanders responsible for the training, recruiting, equipping, and readiness of the Navy Reserve.

Likewise, as a result of its integration efforts, the active Coast Guard assumed responsibility for training the USCGR and assumed evaluation report rating responsibility as well. The USCGR also receives all of its taskings from an active Coast Guard chain of command, and the USCGR was integrated into active Coast Guard day-to-day operations and surge operations as required.

The Army, like the Air Force, is limited in how close it can get to true unity of command in MCUs mixing Title 10 and Title 32 soldiers. In fact, one recent MOA specifies limits to unity of command stating,

> At all times until mobilized, Title 32 Soldiers will be under the command and control of a Title 32 Commander. A limited form of tactical control (TACON) for training may be exercised on the part of the Title 10 Commander, in coordination with the Title 32 Commander who retains administrative control (ADCON) and operational control (OPCON) over the Title 32 soldiers.[5]

[5] Commander, U.S. Army Forces Command; Commander, U.S. Army Reserve; director, U.S. Army National Guard; adjutant general, Wisconsin Army National Guard; and adjutant general, Utah Army National Guard, memorandum of agreement for implementation of the 101st Airborne Division multicomponent unit pilot, January 28, 2016.

Recommendation: Establish Unity of Command to the Greatest Extent Possible in the Integrated Organization

A single, well-defined chain of command is a hallmark of military organizations. Integrating AC and RC organizations can pose some major difficulties in this area, most particularly when integrating Title 10 and Title 32 service members. The cases we examined highlighted the importance of establishing unity of command and the challenges and limitations in doing so. Although the existing legal constraints are unlikely to change much in the near term, the cases also revealed some interesting approaches that can increase unity of command within those constraints.

Address Statutory Barriers

Existing statutes limit the degree of integration and unity of command that can be achieved. There are constraints on the duties that RC members can perform in full-time and part-time roles. The Air Force and the Army face additional statutory constraints as the only services with Title 32 National Guard organizations. Both the Air Force and the Army cases indicate that some integration between AC and National Guard organizations is possible, but it is more difficult to achieve and more limited in nature because of the statutory limitations. For instance, the Air Force's OPDIR approach mentioned in above is an example of a novel approach implemented to achieve as much unity of command as possible across Title 10 and Title 32 components. The Air Force also utilizes the construct of "automatically executing Title 10 orders" as another mechanism to partially overcome the Title 10/Title 32 barrier to integration. When an ANG unit performs Title 10 operational missions, these automatically executing orders facilitate the transfer of personnel from a Title 32 status to a Title 10 status by specifying a clearly identifiable event that activates the transfer to Title 10 status and another clearly identifiable event that terminates the order and transfers the service member back to Title 32 status.

The work-around approaches used by the Air Force and others, such as multiple dual-hats for commanders, suggest that closer integration is possible within existing constraints. However, structural integration of Title 10 and Title 32 forces can only go so far without changes to law and policy. Considering these constraints and ways to work around them to the extent possible is an important practice when undertaking AC/RC integration efforts.

Recommendation: Explicitly Consider Statutory Barriers and Potential Work-Arounds

Title 10 and Title 32 limit the degree of integration and unity of command that can be achieved for the Air Force and the Army. Other statutes limit the functions that RC members can perform. But some work-arounds do exist, and it is important to consider both the limitations and the potential work-arounds before undertaking an integration.

Colocate Active and Reserve Component Personnel in Integrated Organizations

The Army, Coast Guard, and Marine Corps cases highlight the importance of colocating integrated AC and RC units. For instance, in the Army's XVIII Airborne Corps MCU pilot, RC and AC personnel in a MCU are colocated. The corps commander is responsible for the training and readiness of the corps headquarters. The commander has responsibility over both RC and AC personnel. The Army's 101st Airborne Division MCU pilot did not colocate reserve and active personnel, and it did not appear to be as successful as the XVIII Airborne pilot. In his assessment, General McKiernan, the commander of First Army Division East, compared the two models:

> The individual Soldier model implemented by the XVIII Corps integrated staff MCU appears to provide a more effective solution. The non co-located . . . model implemented by the 101st Division did not provide the same flexibility in integration and involved greater challenges with developing and maintaining training plans while managing individual requirements for training days.[6]

In the Coast Guard case, virtually all reserve units are colocated with active units. As a result, the USCGR is integrated into active

---

[6] McKiernan, 2015.

Coast Guard day-to-day operations. This has been made possible by the integrated training of AC and RC personnel. Because RC personnel are now directly assigned to the active units in which they train, their efforts and skills are directly applied to the unit's mission.

The Marine Corps case illustrates the benefits of embedding, and thus colocating, active personnel in RC units and RC personnel in active units. Through its I&I program, the Marine Corps embeds AC personnel in RC units to provide oversight and mentoring. HQMC also has a small number of active reservists embedded in the headquarters staff sections to provide RC perspective as total force policies and actions are developed and executed.

Is Increased Permeability Needed for Active and Reserve Component Integration?

Permeability is described as the ability for people to move between the AC and RC. The extant continuum-of-service policy is an example of this. According to the Army, "The intent of Continuum of Service is to give an individual the opportunity to move back and forth between those different statuses throughout their life and to make that as seamless as possible . . . ."[7] Or, as the Navy describes it, the continuum of service

> provides flexible service options and improves work–life balance, which in turn helps Sailors. Everyone reaches decision points in their careers, and many who serve desire career options other than the "24/7/365 or nothing" proposition. This supports CNO's vision of a seamless Navy Total Force . . . .[8]

In the context of AC/RC integration, the relevant question is whether a lack of permeability is a barrier to integration and, if so, how important a barrier it is. We found no evidence in our literature review or in our case studies that lack of permeability is such a barrier. This is

---

[7] Rob McIlvaine, "Army Planning 'Continuum of Service' Between Components," U.S. Army, November 15, 2011.

[8] Robin R. Braun, vice admiral, U.S. Navy, Chief of Navy Reserve, statement before the U.S. Senate Committee on Appropriations Subcommittee on Defense, April 17, 2013, p. 10.

not to say that permeability would not be a useful enabler for integration. Although it might be a useful enabler for integration and a concept that would be helpful in recruiting and retaining good people, we saw no indication that it is necessary for AC/RC integration to occur.

Bibliography

4th Marine Division Historical Detachment, *History of the 4th Marine Division: 1943–2000*, 2nd ed., 2000.

Alawi, Adel Ismail al-, Nayla Yousif Al-Marzooqi, and Yasmeen Fraidoon Mohammed, "Organizational Culture and Knowledge Sharing: Critical Success Factors," *Journal of Knowledge Management*, Vol. 11, No. 2, 2007, pp. 22–42.

A-Plan Org Chart.pptx, briefing slide, undated.

Beckhard, Richard, and Reuben T. Harris, *Organizational Transitions: Managing Complex Change*, 2nd ed., Reading, Mass.: Addison-Wesley, 1987.

Biddick, Dennis, National Naval Reserve Policy Board, "COMNAVRESFOR Command Brief," briefing slides, April 22, 2016.

Bradley, John A., Gary L. Crone, and David W. Hembroff, "The Next Horizon: Building a Viable Force," *Joint Force Quarterly*, Issue 49, 2nd Quarter 2008.

Brafman, Ori, and Rod A. Beckstrom, *The Starfish and the Spider: The Unstoppable Power of Leaderless Organizations*, New York: Portfolio, 2006.

Braun, Robin R., vice admiral, U.S. Navy, Chief of Navy Reserve, statement before the U.S. Senate Committee on Appropriations Subcommittee on Defense, April 17, 2013.

Brinkerhoff, John R., and Stanley A. Horowitz, *Active–Reserve Integration in the Coast Guard*, Alexandria, Va.: Institute for Defense Analyses, D-1864, October 1996.

Broome, Lissa Lamkin, John M. Conley, and Kimberly D. Krawiec, "Dangerous Categories: Narratives of Corporate Board Diversity," *North Carolina Law Review*, Vol. 89, 2011, pp. 760–808.

Buchanan, Bruce, II, "Government Managers, Business Executives, and Organizational Commitment," *Public Administration Review*, Vol. 34, No. 4, July–August 1974, pp. 339–347.

Burke, W. Warner, and George H. Litwin, "A Causal Model of Organizational Performance and Change," *Journal of Management*, Vol. 18, No. 3, 1992, pp. 523–545.

Burton, Jefferson S., adjutant general, Utah Army National Guard; Donald P. Dunbar, adjutant general, Wisconsin Army National Guard; Timothy J. Kadavy, director, U.S. Army National Guard; Jeffrey W. Talley, commander, U.S. Army Reserve; and Robert B. Abrams, commander, U.S. Army Forces Command, memorandum of agreement for implementation of the 101st Airborne Division multiple-component unit pilot, January 28, 2016.

Caiazza, Amy, "Does Women's Representation in Elected Office Lead to Women-Friendly Policy? Analysis of State-Level Data," *Women and Politics*, Vol. 26, No. 1, 2004, pp. 35–70.

Cantwell, Gerald T., *Citizen Airmen: A History of the Air Force Reserve, 1946–1994*, Air Force History and Museums Program, 1997. As of May 29, 2017: https://media.defense.gov/2010/Dec/01/2001329894/-1/-1/0/AFD-101201-044.pdf

Cartwright, Susan, and Cary L. Cooper, "The Role of Culture Compatibility in Successful Organizational Marriage," *Academy of Management Executive*, Vol. 7, No. 2, May 1993, pp. 57–70.

Chandler, Alfred D., *Strategy and Structure: Chapters in the History of the Industrial Enterprise*, Cambridge, Mass.: MIT Press, 1962.

Chaney, Paul, "Critical Mass, Deliberation and the Substantive Representation of Women: Evidence from the UK's Devolution Programme," *Political Studies*, Vol. 54, No. 4, December 2006, pp. 691–714.

Chief of Naval Operations, *Navy's Total Force Vision for the 21st Century*, Washington, D.C.: U.S. Department of the Navy, January 2010. As of May 26, 2017:
http://www.navy.mil/features/NTF%20Vision%20%28Final%29%2811%20Jan%2010%201210hrs%29.pdf

Childs, Sarah, Paul Webb, and Sally Marthaler, "Constituting and Substantively Representing Women: Applying New Approaches to a UK Case Study," *Politics and Gender*, Vol. 6, No. 2, June 2010, pp. 199–223.

Civil Air Patrol, "Online Media Kit," undated.

Colorado National Guard, "Aerospace Control Alert Mission," November 2015.

Commandant of the Marine Corps, "Marine Requirements Oversight Council," Memorandum 1-02, January 17, 2002. As of May 30, 2017:
http://www.secnav.navy.mil/rda/Policy/2002%20Policy%20Memoranda/cmcmroc10217jan2002.pdf

———, *Total Force Structure Process*, Marine Corps Order 5311.1E, November 18, 2015. As of May 27, 2017:
http://www.marines.mil/News/Publications/ELECTRONIC-LIBRARY/Electronic-Library-Display/Article/900533/mco-53111e/

Commander, FORSCOM—*See* Commander, U.S. Army Forces Command.

Commander, U.S. Army Forces Command, and CAR—*See* Commander, U.S. Army Forces Command, and Chief of the Army Reserve.

Commander, U.S. Army Forces Command, and Chief of the Army Reserve, "Implementation of the XVIII Airborne Corps Multiple-Component Unit (MCU) Pilot," memorandum of agreement, May 8, 2015.

Commander, U.S. Army Forces Command; commander, U.S. Army Reserve; director, U.S. Army National Guard; adjutant general, Wisconsin Army National Guard; and adjutant general, Utah Army National Guard, memorandum of agreement for implementation of the 101st Airborne Division multicomponent unit pilot, January 28, 2016.

Comptroller General, *DOD "Total Force Management": Fact or Rhetoric*, Washington, D.C.: U.S. General Accounting Office, FPCD-78-82, January 24, 1979. As of May 26, 2017:
http://www.gao.gov/assets/130/125320.pdf

Conant, K. J., U.S. Marine Corps Reserve, "History and Overview of Reserve Affairs Division," August 26, 2014.

Correll, John T., "Origins of the Total Force," *Air Force Magazine*, February 2011, pp. 94–97. As of May 26, 2017:
http://www.airforcemag.com/MagazineArchive/Pages/2011/February%202011/0211force.aspx

Costa, Luis Almeida, João Amaro de Matos, and Miguel Pina e Cunha, "The Manager as Change Agent: Communication Channels, Timing of Information, and Attitude Change," *International Studies of Management and Organization*, Vol. 33, No. 4, Winter 2003–2004, pp. 65–93.

Creasey, Timothy J., and Robert Stise, eds., *Best Practices in Change Management: 1120 Participants Share Lessons and Best Practices in Change Management*, 9th ed., Loveland, Colo.: Prosci, 2016.

Cummings, Thomas G., and Christopher G. Worley, *Organization Development and Change*, 5th ed., St. Paul, Minn.: West Publishing, 1993.

Dahlerup, Drude, "From a Small to a Large Minority: Women in Scandinavian Politics," *Scandinavian Political Studies*, Vol. 11, No. 4, December 1988, pp. 275–298.

Damanpour, Fariborz, "Organizational Innovation: A Meta-Analysis of Effects of Determinants and Moderators," *Academy of Management Journal*, Vol. 34, No. 3, September 1991, pp. 555–590.

DCMS—*See* Deputy Commandant for Mission Support.

De Noble, Alex F., Loren T. Gustafson, and Michael Hergert, "Planning for Post-Merger Integration: Eight Lessons for Merger Success," *Long Range Planning*, Vol. 21, No. 4, August 1988, pp. 82–85.

deLeon, Peter, "The Stages Approach to the Policy Process: What Has It Done? Where Is It Going?" in Paul A. Sabatier, ed., *Theories of the Policy Process*, Boulder, Colo.: Westview Press, 1999, pp. 19–34.

Deputy Commandant for Mission Support, U.S. Coast Guard, U.S. Department of Homeland Security, "Reserve and Military Personnel Directorate (CG-13)," undated. As of June 21, 2017:
http://www.dcms.uscg.mil/Our-Organization/
Assistant-Commandant-for-Human-Resources-CG-1/
Reserve-and-Military-Personnel/

———, organizational chart for the Office of the Assistant Commandant for Human Resources, updated October 17, 2016. As of September 26, 2016:
http://www.dcms.uscg.mil/Portals/10/CG-1/docs/pdf/
CG-1_org_chart.pdf?ver=2017-03-20-132345-547

Director of Administration and Management, *Functions of the Department of Defense and Its Major Components*, Department of Defense Directive 5100.01, December 21, 2010. As of May 26, 2017:
http://www.dtic.mil/whs/directives/corres/pdf/510001p.pdf

Drucker, Peter, "The Five Rules of Successful Acquisition," *Wall Street Journal*, October 15, 1981.

Epstein, Marc J., "The Drivers of Success in Post-Merger Integration," *Organizational Dynamics*, Vol. 33, Issue 2, May 2004, pp. 174–189.

Federal Research Division, Library of Congress, *Historical Attempts to Reorganize the Reserve Components*, Washington, D.C., October 2007. As of May 26, 2017:
https://www.loc.gov/rr/frd/pdf-files/
CNGR_Reorganization-Reserve-Components.pdf

Fernandez, Sergio, and Hal G. Rainey, "Managing Successful Organizational Change in the Public Sector," *Public Administration Review*, Vol. 66, No. 2, March 2006, pp. 168–176.

Fleming, Jacqueline, *Blacks in College: A Comparative Study of Students' Success in Black and in White Institutions*, San Francisco, Calif.: Jossey-Bass Publishers, 1984.

Frączkiewicz-Wronka, Aldona, Jacek Szołtysek, and Maria Kotas, "Key Success Factors of Social Services Organizations in the Public Sector," *Management*, Vol. 16, No. 2, December 2012, pp. 231–255.

GAO—*See* U.S. General Accounting Office.

Gardiner, Penny, and Peter Whiting, "Success Factors in Learning Organizations: An Empirical Study," *Industrial and Commercial Training*, Vol. 29, No. 2, 1997, pp. 41–48.

Goman, Carol Kinsey, "The Biggest Mistakes in Managing Change," *Innovative Leader*, Vol. 9, No. 12, December 2000.

Greed, Clara, "Women in the Construction Profession: Achieving Critical Mass," *Gender, Work and Organization*, Vol. 7, No. 3, July 2000, pp. 181–196.

Grey, Sandra, *Does Size Matter? Critical Mass and Women MPs in New-Zealand House of Representatives*, paper written for the 51st Political Studies Association Conference, Manchester, UK, April 10–12, 2001.

Grobman, Gary M., "Complexity Theory: A New Way to Look at Organizational Change," *Public Administration Quarterly*, Vol. 29, No. 3–4, Fall 2005–Winter 2006, pp. 350–382.

Hagedorn, Linda Serra, Winny Chi, Rita M. Cepeda, and Melissa McLain, "An Investigation of Critical Mass: The Role of Latino Representation in the Success of Urban Community College Students," *Research in Higher Education*, Vol. 48, No. 1, February 2007, pp. 73–91.

Hjern, Benny, and David O. Porter, "Implementation Structures: A New Unit of Administrative Analysis," *Organization Studies*, Vol. 2, No. 1, 1981, pp. 211–227.

Hoewing, Gerald L., vice admiral, Deputy Chief of Naval Operations for Manpower, Personnel, Training, and Education, U.S. Navy; and John G. Cotton, vice admiral, Chief of Navy Reserve, U.S. Navy, joint statement before the U.S. House of Representatives Committee on Armed Services Subcommittee on Military Personnel on recruiting and retention, July 19, 2005.

Holt, Daniel T., Achilles A. Armenakis, Hubert S. Feild, and Stanley G. Harris, "Readiness for Organizational Change: The Systematic Development of a Scale," *Journal of Applied Behavioral Science*, Vol. 43, No. 2, June 2007, pp. 232–255.

Jacoby, GEN Charles H., Jr., commander, U.S. Northern Command and North American Aerospace Defense Command, testimony before the National Commission on the Structure of the Air Force, September 26, 2013.

Joint Chiefs of Staff, *Joint Operations*, Joint Publication 3-0, August 11, 2011.

———, *Capstone Concept for Joint Operations: Joint Force 2020*, September 10, 2012. As of May 26, 2017:
http://www.dtic.mil/doctrine/concepts/ccjo_jointforce2020.pdf

———, *Doctrine for the Armed Forces of the United States*, Joint Publication 1, March 25, 2013. As of May 26, 2017:
http://www.dtic.mil/doctrine/new_pubs/jp1.pdf

———, *DoD Dictionary of Military and Associated Terms*, Joint Publication 1-02, November 8, 2010, as amended through February 15, 2016.

Jones, Oswald, "Developing Absorptive Capacity in Mature Organizations: The Change Agent's Role," *Management Learning*, Vol. 37, No. 3, 2006, pp. 355–376.

JP 1-02—*See* Joint Chiefs of Staff, 2010 (2016).

Kanter, Rosabeth Moss, "Some Effects of Proportions on Group Life: Skewed Sex Ratios and Responses to Token Women," *American Journal of Sociology*, Vol. 82, No. 5, March 1977, pp. 965–990.

Keith, Steve, "From the Editor," *Naval Reserve Association News*, Vol. 51, No. 5, May 2004, p. ii.

Konrad, Alison M., Vicki Kramer, and Sumru Erkut, "Critical Mass: The Impact of Three or More Women on Corporate Boards," *Organizational Dynamics*, Vol. 37, No. 2, 2008, pp. 145–164.

Koontz, Christopher N., *Department of the Army Historical Summary: Fiscal Year 2001*, Washington, D.C.: U.S. Army Center of Military History Publication 101-32-1, last updated November 17, 2011. As of May 30, 2017:
http://www.history.army.mil/html/books/101/101-31-2/index.html

Kotter, John P., *A Force for Change: How Leadership Differs from Management*, New York: Free Press, 1990.

———, *Leading Change*, Boston, Mass.: Harvard Business School Press, 1996.

Laird, Melvin R., Secretary of Defense, "Support for Guard and Reserve Forces," memorandum, Washington, D.C., August 21, 1970.

———, *National Security Strategy of Realistic Deterrence: Secretary of Defense Melvin R. Laird's Annual Defense Department Report FY 1973*, February 22, 1972. As of May 27, 2017:
http://www.dtic.mil/dtic/tr/fulltext/u2/a082934.pdf

Lambright, W. Henry, "Leadership and Change at NASA: Sean O'Keefe as Administrator," *Public Administration Review*, March–April 2008, pp. 230–240.

Lipsky, Michael, *Street-Level Bureaucracy: Dilemmas of the Individual in Public Services*, New York: Russell Sage Foundation, 1980.

Lord, Charles G., and Delia S. Saenz, "Memory Deficits and Memory Surfeits: Differential Cognitive Consequences of Tokenism for Tokens and Observers," *Journal of Personality and Social Psychology*, Vol. 49, No. 4, 1985, pp. 918–926.

Marine Corps Order 5311.1E—*See* Commandant of the Marine Corps, 2015.

Marr, Bernard, *Managing and Delivering Performance: How Government, Public Sector, and Not-for-Profit Organizations Can Measure and Manage What Really Matters*, Amsterdam: Butterworth-Heinemann/Elsevier, 2009.

McHugh, John M., Secretary of the Army, "Army Directive 2012-08 (Army Total Force Policy)," memorandum for principal officials of Headquarters, Department of the Army; commanders, U.S. Army Forces Command, U.S. Army Training and Doctrine Command, U.S. Army Materiel Command, U.S. Army Europe, U.S. Army Central, U.S. Army North, U.S. Army South, U.S. Army Pacific, U.S. Army Africa, U.S. Army Special Operations Command, Military Surface Deployment and Distribution Command, U.S. Army Space and Missile Defense Command/ Army Strategic Command, U.S. Army Network Enterprise Technology Command/9th Signal Command (Army), U.S. Army Medical Command, U.S. Army Intelligence and Security Command, U.S. Army Criminal Investigation Command, U.S. Army Corps of Engineers, U.S. Army Military District of Washington, U.S. Army Test and Evaluation Command, and U.S. Army Installation Management Command; superintendent, U.S. Military Academy; and director, U.S. Army Acquisition Support Center, Washington, D.C.: U.S. Department of Army, September 4, 2012. As of May 26, 2017: http://www.apd.army.mil/epubs/DR_pubs/DR_a/pdf/web/ad2012_08.pdf

McIlvaine, Rob, "Army Planning 'Continuum of Service' Between Components," U.S. Army, November 15, 2011. As of May 31, 2017: http://www.army.mil/article/69397

McKiernan, Brian J., major general, U.S. Army, "Multi-Component Unit (MCU) Final Findings and Recommendations," memorandum for commanding general, U.S. Army Forces Command, December 18, 2015.

Mechanic, David, "Sources of Power of Lower Participants in Complex Organizations," *Administrative Science Quarterly*, Vol. 7, No. 3, December 1962, pp. 349–364.

Meier, Kenneth J., and Laurence J. O'Toole Jr., "Public Management and Organizational Performance: The Effect of Managerial Quality," *Journal of Policy Analysis and Management*, Vol. 21, No. 4, Autumn 2002, pp. 629–643.

Mills, Richard P., lieutenant general, U.S. Marine Corps, commander, "Vision of the Commander, Marine Forces Reserve," undated. As of May 30, 2017: http://www.marforres.marines.mil/Portals/116/Docs/CmdDeck/ VisionCMFR.JPG

Moeller, Michael R., deputy chief of staff, Strategic Plans and Programs, U.S. Air Force, "Air Force Guidance Memorandum to AFI 90-1001," Washington, D.C., Air Force Guidance Memorandum 01 to Air Force Instruction 90-1001, January 23, 2014.

Moon, Michael, "Bottom-Up Instigated Organizational Change Through Constructionist Conversation," *Journal of Knowledge Management Practice*, Vol. 9, No. 4, December 2008.

National Commission on the Future of the Army, *Report to the President and the Congress of the United States*, January 28, 2016. As of May 26, 2017: http://www.ncfa.ncr.gov/content/download-full-report.html

National Commission on the Structure of the Air Force, *Report to the President and Congress of the United States*, Washington, D.C., January 30, 2014. As of May 26, 2017:
http://policy.defense.gov/Portals/11/Documents/hdasa/AFForceStructureCommissionReport01302014.pdf

Navas, William A., Jr., "Integration of the Active and Reserve Navy: A Case for Transformational Change," *Naval Reserve Association News*, Vol. 51, No. 5, May 2004, pp. 11–19.

NCFA—*See* National Commission on the Future of the Army.

NCSAF—*See* National Commission on the Structure of the Air Force.

Nofi, Albert A., *The Naval Militia: A Neglected Asset?* Alexandria, Va.: CNA Corporation, CIM D0015586.A1/Final, July 2007. As of May 26, 2017:
https://www.cna.org/CNA_files/PDF/D0015586.A1.pdf

Office of the Assistant Secretary of Defense for Manpower and Reserve Affairs, Total Force Planning and Requirements Directorate, *Defense Manpower Requirements Report: Fiscal Year 2017*, Washington, D.C.: U.S. Department of Defense, April 2016. As of May 26, 2017:
http://www.people.mil/Portals/56/Documents/tfprq/FY17%20DMRR%20Final.pdf

Pascale, Richard T., and Jerry Sternin, "Your Company's Secret Change Agents," *Harvard Business Review*, May 2005. As of May 26, 2017:
https://hbr.org/2005/05/your-companys-secret-change-agents

Pettigrew, Andrew M., Richard W. Woodman, and Kim S. Cameron, "Studying Organizational Change and Development: Challenges for Future Research," *Academy of Management Journal*, Vol. 44, No. 4, August 2001, pp. 697–713.

Pfeffer, Jeffrey, and John F. Veiga, "Putting People First for Organizational Success," *Academy of Management Executive*, Vol. 13, No. 2, May 1999, pp. 37–48.

Pilling, Donald L., vice chief of naval operations, "Memorandum on OPNAV Instruction 1001.21B to All Ships and Stations (Less Marine Corps Field Addresses Not Having Navy Personnel Attached)," Washington, D.C., 1998.

Plewes, Thomas J., chief, Army Reserve, U.S. Army, "HQDA Realignment Implementation Plan—Office, Chief Army Reserve," memorandum for the Deputy Under Secretary of the Army for International Affairs, November 3, 2001.

Public Law 68-512, An Act to Provide for the Creation, Organization, Administration, and Maintenance of a Naval Reserve and a Marine Corps Reserve, February 28, 1925.

Public Law 112-239, National Defense Authorization Act for Fiscal Year 2013, January 2, 2013. As of May 29, 2017:
https://www.gpo.gov/fdsys/pkg/PLAW-112publ239/content-detail.html

Purchase, Bryne, "Strategies for Implementing Organizational Change in a Public Sector Context: The Case of Canada," *TDRI Quarterly Review*, Vol. 11, No. 4, December 1996, pp. 27–35.

Rangone, Andrea, "Linking Organizational Effectiveness, Key Success Factors and Performance Measures: An Analytical Framework," *Management Accounting Research*, Vol. 8, No. 2, June 1997, pp. 207–219.

Reddish, Chris, Forces Command, and Mark Berglund, National Guard Bureau, briefing on multicomponent units to the National Commission on the Future of the Army, Washington, D.C., August 17, 2015.

Reserve Forces Policy Board, *Eliminating Major Gaps in DoD Data on the Fully-Burdened and Life-Cycle Cost of Military Personnel: Cost Elements Should Be Mandated by Policy—Final Report to the Secretary of Defense*, U.S. Department of Defense, Report FY13-02, January 7, 2013. As of May 26, 2017:
http://rfpb.defense.gov/Portals/67/Documents/
RFPB_Cost_Methodology_Final_Report_7Jan13.pdf

Richman, Laura Smart, Michelle vanDellen, and Wendy Wood, "How Women Cope: Being a Numerical Minority in a Male-Dominated Profession," *Journal of Social Issues*, Vol. 67, No. 3, September 2011, pp. 492–509.

Robbert, Al, William A. Williams, and Cynthia R. Cook, *Principles for Determining the Air Force Active/Reserve Mix*, Santa Monica, Calif.: RAND Corporation, MR-1091-AF, 1999. As of May 28, 2017:
https://www.rand.org/pubs/monograph_reports/MR1091.html

Rostker, Bernard, Charles Robert Roll Jr., Marney Peet, Marygail Brauner, Harry J. Thie, Roger Allen Brown, Glenn A. Gotz, Steve Drezner, Bruce W. Don, Ken Watman, Michael G. Shanley, Fred L. Frostic, Colin O. Halvorson, Norman T. O'Meara, Jeanne M. Jarvaise, Robert Howe, David A. Shlapak, William Schwabe, Adele Palmer, James H. Bigelow, Joseph G. Bolten, Deena Dizengoff, Jennifer H. Kawata, Hugh G. Massey, Robert Petruschell, Craig Moore, Thomas F. Lippiatt, Ronald E. Sortor, J. Michael Polich, David W. Grissmer, Sheila Nataraj Kirby, and Richard Buddin, *Assessing the Structure and Mix of Future Active and Reserve Forces: Final Report to the Secretary of Defense*, Santa Monica, Calif.: RAND Corporation, MR-140-1-OSD, 1992. As of May 26, 2017:
https://www.rand.org/pubs/monograph_reports/MR140-1.html

Ryan, Andrew, deputy, strategic communications officer, Office of Marine Forces Reserve, "Total Force Integration Brief," briefing slides from briefing to MG (ret.) G. A. Schumacher, U.S. Army, last updated June 28, 2016.

Sabatier, Paul A., "Top-Down and Bottom-Up Approaches to Implementation Research: A Critical Analysis and Suggested Synthesis," *Journal of Public Policy*, Vol. 6, No. 1, January–March 1986, pp. 21–48.

Sanger, Mary Bryna, "Does Measuring Performance Lead to Better Performance?" *Journal of Policy Analysis and Management*, Vol. 32, No. 1, Winter 2013, pp. 185–203.

Santelli, James S., *A Brief History of the 4th Marines*, Washington, D.C.: Historical Division, Headquarters, U.S. Marine Corps, Marine Corps Historical Reference Pamphlet, 1970. As of May 30, 2017:
http://www.marines.mil/Portals/59/Publications/A%20Brief%20History%20of%20the%204th%20Marines%20%20PCN%2019000245500_1.pdf?ver=2012-10-11-163144-597

Sawhill, John C., and David Williamson, "Mission Impossible? Measuring Success in Nonprofit Organizations," *Nonprofit Management and Leadership*, Vol. 11, No. 3, Spring 2001, pp. 371–386.

Schaefer, Agnes Gereben, Jennie W. Wenger, Jennifer Kavanagh, Jonathan P. Wong, Gillian S. Oak, Thomas E. Trail, and Todd Nichols, *Implications of Integrating Women into the Marine Corps Infantry*, Santa Monica, Calif.: RAND Corporation, RR-1103-USMC, 2015. As of May 26, 2017:
https://www.rand.org/pubs/research_reports/RR1103.html

Schneider, Benjamin, Sarah K. Gunnarson, and Kathryn Niles-Jolly, "Creating the Climate and Culture of Success," *Organizational Dynamics*, Vol. 23, No. 1, Summer 1994, pp. 17–29.

Secretary of the Army, "Designation of Associated Units in Support of Army Total Force Policy," memorandum for principal officials of Headquarters, Department of the Army, and commanders, U.S. Army Forces Command, U.S. Army Training and Doctrine Command, and U.S. Army Pacific, March 21, 2016.

Sekaquaptewa, Denise, and Mischa Thompson, "Solo Status, Stereotype Threat, and Performance Expectancies: Their Effects on Women's Performance," *Journal of Experimental Social Psychology*, Vol. 39, No. 1, January 2003, pp. 68–74.

Shinseki, Eric K., chief of staff, U.S. Army, and Thomas E. White, Secretary of the Army, "Implementation Plan for Realignment," memorandum for the Chief of Army Reserve of the Army, January 4, 2002.

Talley, Jeffrey W., chief, Army Reserve, U.S. Army, and Mark A. Milley, commander, U.S. Army Forces Command, "Implementation of the XVIII Airborne Corps Multiple-Component Unit (MCU) Pilot," memorandum of agreement, May 8, 2015.

"The Integration of Active and Reserve Forces," *Coast Guard Reservist*, October 1994.

Total Force Policy Study Group, *Total Force Policy Interim Report to the Congress*, U.S. Department of Defense, September 1990. As of May 26, 2017:
http://www.dtic.mil/get-tr-doc/pdf?AD=ADA235382

U.S. Air Force, "Air Force Continues to Pursue Total Force Integration," press release, March 11, 2016. As of May 26, 2017:
http://www.af.mil/News/Article-Display/Article/691270/
air-force-continues-to-pursue-total-force-integration/

USCGR—*See* U.S. Coast Guard Reserve.

U.S. Coast Guard, *U.S. Coast Guard Overview*, Washington, D.C.: U.S. Coast Guard Headquarters, CG-PTT, October 2016. As of August 27, 2017:
http://www.overview.uscg.mil/Portals/6/Documents/PDF/
USCG_Overview.pdf?ver=2016-10-21-114442-890

———, "Units," last modified July 5, 2017. As of June 21, 2017:
https://www.uscg.mil/top/units/default.asp

U.S. Coast Guard Auxiliary, "About the Auxiliary," last updated July 14, 2015. As of May 30, 2017:
http://cgaux.org/about.php

U.S. Coast Guard Reserve, "Workforce Organization," undated. As of August 28, 2017:
http://www.reserve.uscg.mil/Workforce-Organization/

U.S. Code, Title 10, Armed Forces, Subtitle E, Reserve Components, Part I, Organization and Administration, Chapter 1003, Reserve Components Generally, Section 10101, Reserve Components Named. As of May 26, 2017:
https://www.gpo.gov/fdsys/granule/USCODE-2011-title10/
USCODE-2011-title10-subtitleE-partI-chap1003-sec10101

U.S. General Accounting Office, *Comptroller General's Annual Report 1979*, Washington, D.C., B-119600, January 25, 1980. As of May 26, 2017:
http://www.gao.gov/products/111905

———, *Force Structure: Army Is Integrating Active and Reserve Combat Forces, but Challenges Remain*, Washington, D.C., GAO/NSIAD-00-162, July 2000. As of May 30, 2017:
http://www.gao.gov/products/NSIAD-00-162

U.S. Marine Corps, "Headquarters Marine Corps," home page, undated. As of February 17, 2017:
http://www.hqmc.marines.mil/

van de Ven, Andrew H., and Marshall Scott Poole, "Explaining Development and Change in Organizations," *Academy of Management Review*, Vol. 20, No. 3, July 1995, pp. 510–540.

Volesky, Gary J., major general, U.S. Army, commanding general of the 101st Airborne Division (Air Assault) and Fort Campbell, "101st Airborne Division (Air Assault) Multi-Component Unit (MCU) Assessment of Mission Effectiveness," memorandum commanding general, U.S. Army Forces Command, December 16, 2015.

Wang, Jia, "Applying Western Organization Development in China: Lessons from a Case of Success," *Journal of European Industrial Training*, Vol. 34, No. 1, 2010, pp. 54–69.

CPSIA information can be obtained
at www.ICGtesting.com
Printed in the USA
LVHW041658231019
635121LV00014B/222/P